Bad Call

Inside Technology
edited by Wiebe E. Bijker, W. Bernard Carlson, and Trevor Pinch

Bad Call

**Technology's Attack on Referees and Umpires
and How to Fix It**

Harry Collins, Robert Evans, and Christopher Higgins

The MIT Press
Cambridge, Massachusetts
London, England

This book was set in Stone Sans and Stone Serif by Toppan Best-set Premedia Limited. Printed and bound in the United States of America.

Library of Congress Cataloging-in-Publication Data

Names: Collins, H. M. (Harry M.), 1943- author.
Title: Bad call : technology's attack on referees and umpires and how to fix it / Harry Collins, Robert Evans, and Christopher Higgins.
Description: Cambridge, MA : The MIT Press, [2016] | Series: Inside technology | Includes bibliographical references and index.
Identifiers: LCCN 2016013520 | ISBN 9780262035392 (hardcover : alk. paper)
Subjects: LCSH: Sports officiating. | Sports--Technological innovations. | Soccer--Officiating. | Cricket--Umpiring.
Classification: LCC GV735 .C65 2016 | DDC 796--dc23 LC record available at https://lccn.loc.gov/2016013520

10   9   8   7   6   5   4   3   2   1

Contents

Introduction

"Yeeeeeeeeeeeesss!" That's me, Harry Collins, a seventy-year-old university professor, my thirty-eight-year-old son, who works for an economics think tank, and my thirty-five-year-old daughter, a senior civil servant. We're high-fiving and rolling about on the floor. The final whistle has just gone and Liverpool has held on at Anfield, their home ground, to beat Manchester City and put the Premiership title in their own hands. It's Sunday, April 13, 2013, and we've been watching the telly at a friend's house in London, where my kids live. My wife has gone home to Cardiff on the train because she can't stand the tension and she doesn't want to see me have a heart attack. Insanity!

But Manchester City fans must be, as the football community says, "gutted."[1] Not only because they've lost, but because they've been cheated. There should have been three penalties in that game, and none of them were given. City's Vincent Kompany put two hands around Liverpool's Luis Suárez's chest and pushed him to the ground in the penalty area; Liverpool's Mamadou Sakho took a kick at the ball and missed it, hitting a City player instead; and Liverpool's Martin Škrtel simply punched an incoming ball out of the penalty area with his fist even though he's not the goalkeeper and is not allowed to touch it with his

Figure 0.1
Manchester City 2, Liverpool 1, December 26, 2013. Raheem Sterling is "miles" onside as the ball is passed to him; he was flagged offside.

hands. If those three penalties had been given, the match would have been a draw and City would still have had the Premiership title in their hands.

Are we saddened by the fact that we won by foul means? No—we're whooping! First, because football fans like to win however it is done. But more importantly because this balances out what happened earlier in the season, when we played City at their stadium in Manchester. Liverpool's Raheem Sterling was through on goal and would surely have scored but was flagged "offside" when he was two yards onside. City won, when it should have been a draw. What we saw on this occasion was the play of chance giving us justice. We should have had two more

Figure 0.2
Martin Škrtel punches the ball out of the Liverpool penalty area. It was
not spotted by the referee.

points at the Etihad, and City should have had two more points
at Anfield. As they say, "It all balances out in the end."

At least, that's what most people say, but it isn't true; and
this book will prove it. What actually happens is that football
fans are cheated, week after week. It is not only football fans,
of course. On June 2, 2010, Armando Galarraga of the Detroit
Tigers pitched a nearly perfect baseball game (it would have been
only the twenty-first ever at this level) that was spoiled by an
incorrect call, obvious to TV viewers and still obvious on the TV
replays accessible on the Web; the runner was clearly out but
the umpire called him safe (see https://en.wikipedia.org/wiki/
Armando_Galarraga%27s_near-perfect_game). The umpire even
admitted, "I just cost that kid a perfect game." But the baseball

authorities learned their lesson, and since 2014 instant TV replays are available to baseball umpires. We will argue among many other things that they should be introduced in football.

Going back to football, we say fans are cheated because the lung-bursting effort and gut-wrenching emotion they imaginatively share with their teams is, as in the case of the Galarraga incident, made futile by clearly mistaken refereeing decisions. Ten million hearts are broken every week, and they do not need to be. "Cheated" is a strong word. The referees are not cheating; they are doing their best, and they often suffer along with the fans. Jim Joyce, the umpire who made the wrong call in the Galarraga incident described above, immediately felt terrible. In the book he wrote with Galarraga, he remarks that even though his mistake was not shown on a big screen, just a few moments later he felt humiliated as the fans were watching it on hand-held devices and smaller monitors around the stadium (Galarraga and Joyce 2012, 212). A little after the game in conversation with reporters, he writes that he said "I feel like I took something away from that kid and I don't know how to give it back." And then he writes, "The rest is a bit of a blur because I break down at this point. Right then and there I start crying" (217). The players may cheat from time to time, but they too are not cheating the fans; they are trying to do their best for them.

Rather, it is the football authorities who are cheating the fans, because they could easily fix the problem but they won't. In this book we'll explain how they could easily fix it. We'll ask why there has been such a fuss over the introduction of "goal-line technology," used to determine whether a ball has completely crossed the goal line, when, in the 2011–12 and 2012–13 Premiership seasons, as far as we can see from the work presented in chapters 5 and 7, it could have corrected only five mistakes in total, and in the 2013–14 season, when it was brought into use, it resolved only six doubtful cases. That's a maximum of eleven cases in three seasons. In those same three seasons we will be asking why, given our way of assessing these things, nothing was done about 161 incorrect penalty decisions, 86 incorrect offside decisions where goals were at stake, and 88 wrong decisions regarding red cards followed by sending offs—all crucial to the outcome of the games. Of course, you will have to decide for yourself if our way of assessing things is right, but even if it's "miles out" it still makes goal-line technology pretty well irrelevant, while almost the entirety of what is going wrong can be fixed quite simply. Positions in the league table can change based on a point or even a single goal, so we'll ask why, based on our calculations, individual clubs in the Premiership were cheated of up to a net nine points in the 2011–12 season or were unfairly awarded up to ten, cheated of up to seven points in 2012–13 and unfairly awarded up to six, and in the 2013–14 season were cheated of up to nine points and unfairly awarded up to eight. We'll show that if these wrong decisions had been corrected, different clubs would have won the Premiership, different clubs would have filled the Champions League places, and different clubs would have found themselves in the relegation zone. So things didn't "all balance out in the end." And we'll show how easy it would be to put all this right and let the players' skill and effort decide the outcome of football matches rather than referees' bad calls.

The results we've just outlined will be found chapters 6 and 7, but first, in chapters 1 and 2, we look at refereeing and umpiring in a new way and present a different way to think about the various technologies developed to aid umpires and referees. It is only by pulling all this together that we can justify our recommendation, which, to anticipate, is that TV replays should be regularly used by football referees for all the kinds of incidents described above, including goal-line disputes, while the way technologies are used in most other sports should be changed. We suggest that wherever possible, complicated technology should be avoided: "Use only what you need and nothing overly complicated or difficult to understand." When complicated technologies are used, such as the Hawk-Eye "reconstructed track device" and its counterparts made by other firms, they should be used carefully to help match officials, not to replace them. "Reconstructed track devices," by the way, is a bit of a mouthful, so from here on we'll refer to them as "track estimators." Track estimators use television cameras to try to reconstruct the track of the ball in relationship to the field of play and to try to predict its future path, impact point, and so forth. As we will see, even tracking a ball and estimating its impact point involves a certain amount of forward prediction from past data. We will argue that, currently, track estimators are used well in cricket but poorly in tennis, where they present a misleading impression of perfect accuracy.

We now set out the main arguments of the book that lead up to the reexamination of track estimators and to the three years of Premiership football. Among other things, we want to explain how this book differs from others that analyze decision making in sports. First, this book focuses on those who watch sports mostly on their home television screens which provide continual replays of incidents, and those who access TV replays on the big screens at some stadiums. This book is for television viewers, referees, and umpires, and it may be the first book on sports to put the needs of the TV viewer right up there with the other people watching, playing, officiating, and administrating.

Second, the book blends a number of different ways of looking at things—we start, chapter 1, with some easy-to-understand
philosophy to explain the nature of umpiring and refereeing. We
then combine this with an understanding of the technologies
used in sports to help match officials (chapter 2) and a simple
explanation of the way measuring is done in science and engineering (chapters 3–5). It may seem complicated, but you need
all these different ways of looking at things to understand how
to solve the problem of better refereeing and better TV viewing.
Without the philosophy it is too easy to be blinded by measurement; without the understanding of measurement it is too easy
to be blinded by the technology; without the understanding of
technology it is too easy to think that more technology is always
better. What we conclude, and what drives the whole book forward, is that judging what is happening on a sports field is a
matter not of ever-increasing accuracy but of reconciling what
television viewers see and what match officials see—it is a matter
of justice.

Resolving the problem does not require technology that is
any fancier or more complex than the technology that caused
the problem in the first place—regular broadcast-TV cameras.
The drive toward higher and higher technology is motivated by
conceiving of the problem of sports-officiating as exact accuracy
in decision making. Thus, the high technologies claim to be
able to resolve disputes about the position of a ball to within a
millimeter; but such resolution is not needed if no human can
resolve ball-positions in that way. Exact accuracy is, in any case,
a shibboleth; this is obvious in games such as tennis, football, or
rugby where the ball is filled with air. Such balls are "squashy,"
and one cannot know the position of their outside envelope to
within a millimeter because such accuracy does not exist. Nor

are goalposts and the like erected using a micrometer. Again, in any game that uses lines as defining criteria, such as the goal line painted on a grass field in football, to say that one can define its edge to within a millimeter is ridiculous. What happens in the case of the high technologies such as track estimators is that the actual playing area and ball are replaced by a computer-generated virtual playing area and ball, but this is never made clear to the public; rather, the virtual is presented as actual. Nor is the public ever shown how likely it is that line calls produced by track estimators will be in error. In the body of the book we point out a number of such fallacies and misleading impressions.

What matters is not exactly where the edge of the ball is but where it appears to be to the human eye, just as has always been the case in sports. It is just that now the human eye can be enhanced, and is regularly enhanced, by TV replays. The use of this relatively low-level technology can solve the problems without misleading anyone. We show how it can be done without taking ultimate authority away from match officials and without slowing the game much or at all. The answer is that the umpire's decision must stand unless TV can *rapidly* show that it was obviously wrong. If TV cannot rapidly show that a decision is wrong, then TV-watching fans cannot be seeing any obvious injustice.

But before we get to the details, we begin with a discussion of what justice means in sports.

1   Justice and Decision Making in Sports

I call 'em as they are.
I call 'em as I see 'em.
They ain't anything 'til I call them.
—Supposed baseball umpires' competing philosophies in respect of calling "ball" or "strike"

Not only must Justice be done; it must also be seen to be done.
—From a landmark English court ruling of the 1920s

The means to make sound judgments about what happens in sports has become "distributed" to new groups of people, such as TV viewers and commentators. Consequently, TV viewers can now question referees' and umpires' decisions. This questioning can now be done on good grounds because of our access to TV replays.

Making Reality

We will start with some philosophy and the introduction of some new terms. This may seem a bit odd in a book about sports, but the topic of sports is as subject to philosophical analysis as any other human activity. We'd like this bit of philosophy to be among the things discussed in the pub after a game, but whether it is or not, it will pull the whole argument of the book together. So we will explain it as simply as possible.

"Ontology" is a term that frightens even some philosophers, but it is quite simple to explain. Ontology is the philosophy of what exists. If you think, like some of the ancient Greeks, that all substances are made out of water put together in different ways, you are thinking ontologically. If you wonder sometimes if the whole of your experience is really a dream, your wonderings are ontological—you are thinking about what really exists. Bill Klem, the baseball umpire who said "They ain't anything 'til I call them," was making an ontological claim: he was saying that the existence of "balls" and "strikes" depends on his saying they are balls and strikes, and that until an umpire decides whether they are balls and strikes, those balls and strikes do not exist. Klem, whether he knew it or not, was acting as a philosopher in the subfield of ontology when he said balls and strikes come into existence as a result of the umpire's words. The same goes for all umpires and referees who, when faced with an irate player shouting "That wasn't out," respond with "Look in tomorrow's newspaper." Reports in the newspapers and the putting together of the averages and other statistics in the archives collect together the reality that umpires and referees have made— the balls, the strikes, the "ins," the "outs," the penalties, and the offsides. When one has hold of that idea, one can say that a characteristic of umpires and referees is that they have "ontological authority": they are granted the power to create bits of reality. The Galarraga case mentioned in the introduction illustrates this well: even though all parties involved knew shortly after the decision had been made that the umpire had it wrong, and the authorities were requested to change the record and award the pitcher a perfect game, they did not; everyone knew that reality had been made by the umpire and it could not be unmade. Ontological authority is an idea that will run through the book as we look at what umpires and referees do and how they can do it better. It's a forbidding-looking term at first sight, but it's not really that difficult an idea.

"Epistemology" is another complicated word taken from the language of philosophers. This too is not hard to understand. Epistemology is about how we know things. If you say, "The best way to work out whether a salesman is honest is to trust your instinct," you are making an epistemological claim. If you say, "The only way to know anything certain about the natural world is through scientific research," you are in the realm of epistemology. If you say, "I can tell a ball from a strike because I have been watching baseball all my life," you are an epistemologist, without necessarily knowing it.

This puts us in a position to introduce another new term, which we will use over and over: "epistemological privilege." We can say that referees and umpires are granted ontological authority because they have epistemological privilege—which, translated back into ordinary words, means that they are better at judging what is happening than the rest of us because, among other things, they have had a lot of practice.

The source of referees' and umpires' epistemological privilege is actually threefold: they are usually in a better location than the rest of us in respect of what is happening on the field so they have a better view; they usually have to undergo a course of training to refine their understanding of how to make good judgments, where to position themselves on the field in moving ball games, and so forth; and not only are they usually experienced at judging but it is a reasonable expectation that they will be fair and neutral in the application of their epistemological privilege. The two (possibly mythical) baseball umpires who claim "I call 'em as they are" and "I call 'em as I see 'em" are claiming, directly or indirectly, that they are pretty good at seeing 'em as they are. That is, they are pointing out their epistemological privilege.

We begin this chapter with the claims of baseball umpires because, though baseball is not going to be analyzed in great detail in this book, the distinction between balls and strikes in baseball provides a very good illustration of the philosophy of match officiating. The strike zone is wonderfully ill-defined: there are no markings and it is delimited in space in relationship to the dimensions of the batter. We understand that its upper bound has moved considerably over the years as a result of changing fashions among umpires and the demand from fans for more hits and fewer "outs." We understand that technology has been introduced to aid in the definition of the strike zone, and we hope that someone will analyze the use of that technology by applying some of the ideas and terms introduced in this book.

Returning to the main theme, in the pre-TV era match officials' ontological authority meshed smoothly with their epistemological privilege, and umpiring and refereeing worked fairly well. New technologies, however, have degraded epistemological privilege. Where a TV screen is present in a stadium, the epistemological privilege has migrated to the crowd; and whenever a match is televised, it has shifted even further to the remote television viewer. But the ontological authority has not shifted— thank goodness; it is not the crowd who decides what exists in terms of penalties, offsides, and sending offs, and that is just as well, as we know that every fan is biased.

Consider the opposite case. A popular Saturday-night show on British TV is *Strictly Come Dancing* (the US version is *Dancing with the Stars*). On this show, celebrities are taught ballroom dancing by professional dancing partners and compete with other celebrities in front of a panel of expert judges. The judges give their scores, and each week one couple—celebrity plus professional partner—is eliminated. But the public also has a say. The public can text in votes or use the Internet to say who they think should remain in the competition and who should be eliminated. The result is often that a popular celebrity will be saved from elimination even though he or she is a palpably less skillful dancer than the person who is thrown out; one sees some of the judges wilt in the face of these decisions that have little to do with dancing ability. In our language, the epistemological privilege has stayed with the judges but the ontological authority has shifted to the crowd: just the opposite of what happens in the case of TV replays of sporting events (and, if anything significant depended on it, far worse). Perhaps one can begin to see how useful this philosophical language is: it gives us a clean and clear way of contrasting *Strictly Come Dancing* with sports.

The sports situation might be preferable, but, often, a strong disharmony emerges between the crowd's epistemological privilege, generated by TV replays, and the referee's ontological authority, leading to loss of credibility of the match official and the sport itself, as well as unhappiness among spectators with access to a TV. To describe the sources of this unhappiness we now introduce a new fourfold classification of types of justice: "presumptive justice," "transparent justice," "transparent injustice," and "false transparency." Once more these are all easy to understand.

Justice

Most of us don't spend our days in courtrooms. We read about court cases in the newspapers and see them reported on TV. But we assume that if we had been in the courtroom, and/or followed the police investigations preceding the trial, we would have seen justice being done. Usually we have enough faith in the system to take it that the cases we hear about have been fairly investigated and tried. This is what we will define as *presumptive justice*—it is what we presume takes place in the justice system even though we do not see it happening with our own eyes.

For the most part, sports used to depend on presumptive justice, before the introduction of TV. Sports fans presumed that justice was being done as well as it could be done when the match official was exercising ontological authority. No one was better placed or more capable of doing the job than the referee or umpire—he had the most epistemological privilege—and it was reasonable to presume that if you were in the official's position and possessed the official's skills and experience you would have made the same decisions. Presumptive justice worked pretty well in sports until new technologies came along.

Transparent justice, in contrast, simply means justice that is in fact seen to be done. It is what one hopes one would see if one were a member of a jury in a court case. In sports, transparent justice has always been available regarding some decisions, since the crowd could see the net bulge when hit by the soccer ball, see the bails fall when the wicket was broken by the cricket ball, see the high-jump bar still in position, or not, when the jumper landed on the other side, see the American football player carry the ball into the end zone, and see the trajectory of the home run in baseball. Transparent justice has become more prevalent with television replays. Television has brought certain old instances of traditional presumptive justice into the realm of transparent justice because, given the zoom lens, the TV viewer can now sometimes see for herself what the referee or umpire is seeing when she could not see it before. TV helps the crowd see the baseball runner tagged or home, see the pass interference in American football, see the jumper's foot go over the baseboard, see the cricket fielder throw the ball into the air before over-balancing and stepping across the boundary and then stepping back infield and catch the ball fairly, and see the football player push another in the penalty area as the cross comes in.

But with TV, the opposite of transparent justice is also more prevalent. The opposite is *transparent injustice*. The TV viewers, more and more, see the referee getting things wrong. Transparent injustice rarely happened in traditional sports officiating because no one in the crowd could see the referee's mistakes.

There is one more category of justice to fill in. This is *false transparency*, which we will sometimes refer to as *show-trial justice*. False transparency occurs where justice appears to be being done but is not in fact being done. In the early days of Stalinist show trials, there was false transparency—it looked as though the accused were freely confessing to their crimes and justice was being seen to be done; later, when it was realized that the confessions were extracted by torture and threats to the victim's family, these cases of apparent justice became cases of transparent *in*justice. Fortunately, the only case of sports officiating where we will need to draw on the category of show-trial justice is in the use of what we will call "track estimators"—such as the well-known Hawk-Eye—and then only in tennis.

Armed with these ideas and the terms that capture them, we can now take a look at the different groups who have an interest in referees' and umpires' decisions.

The Players

At the lowest level of games such as cricket, tennis, baseball, or football, there are no umpires or referees. It is almost certain that even those who eventually go on to become professional sports players first played the favored games without any match officials. In these "pickup" games, played on a patch of waste ground or in the street, with a pile of coats for goalposts or wicket, decisions are made by agreement among the players. Such decisions won't always be smooth, but the players will work it out if they want the game to continue.

When we move from a few kids gathered in the street to the level exemplified by playing slightly more organized cricket at (high) school or for one's departmental team at university, a person is selected to act as umpire or referee. In cricket, the umpire is chosen from the members of the batting team who are not batting at that time. In amateur football, referees and their assistants are typically provided by the teams taking part. In both these cases, a remarkable thing has to happen if there is to be any justice on display: the temporary referee or umpire has to act neutrally, or near neutrally, even though he or she belongs to one of the teams taking part. These stand-in match officials seem to be able to manage it, or at least give a pretty good appearance of managing this remarkable feat. It seems almost certain, therefore, that pretty much every player of every sport has, to a greater or lesser extent, shared in the role of match official at some stage of their life. Because, from time to time, they will have had to make decisions that go against the interests of their teammates, they have learned that justice is central to umpiring and refereeing; they will also have learned, in the course of the inevitable arguments, that it is good to be able to demonstrate *that* justice has been done, as well as simply doing it.

This means that players taking part in a professional game are well equipped to be match officials in that they know what it is to be neutral, they have an excellent view of what is going on, and they have good experience of the game they are playing; in sum, they have great epistemological privilege. Therefore, setting aside the fact that they actually have no interest in acting neutrally, it is not surprising that they sometimes feel they have the right to argue with the umpire or referee. Using the philosophical language, there is rough parity between the epistemological privilege of players and umpires even though ontological authority stays with the umpire; and this can lead to some tension. The extent to which players argue with match officials varies from sport to sport—for example, it is not allowed in cricket, leading to severe financial penalties in the professional game—whereas John McEnroe's complaints to tennis umpires have come to be regarded as an art form in themselves. But referees are generally seen to ignore protests, however vehement, as they must in order to retain their ontological authority.

The Crowd

Traditionally, in respect of most decisions, the crowd seated at a sports arena is in a relatively poor viewing position, they will have had little or no referee training, and they are not neutral. They are thus at a sharp epistemological disadvantage compared to the match officials. Traditionally, then, crowds' berating of referees or umpires during a match has generally been recognized as being more a matter of partisanship than superior judgment.

Television Viewers and Commentators

To reiterate, with the introduction of television, any crowd who benefits from replay screens located at the sports venue has lost its epistemological disadvantage with regard to a range of decisions. With multiple cameras being used, often at least one camera angle will provide a vantage point that is as good as, or better than, that of the umpire or referee. Replays, especially slow-motion replays, also put the television viewer, whether at home or at the match, in a still better position to make a judgment than the umpire or referee. The television replay destroys the "superior view" advantage of umpires. In many cases it also destroys their "specialist skill" advantage, since a good part of that specialist skill is to make the right decision in real time. In live officiating, the umpire or referee has to judge an almost instantaneous sequence of events that require processing as much at the unconscious level as at the conscious level. This takes the experience and practice required to grasp a situation as a whole rather than to assemble a decision out of discrete observations. To put this another way, part of specialist umpiring and refereeing skills comprises "somatic tacit knowledge"—the kind of skills we build up in our bodies like the ability to drive, or type fluently, or ride a bike without thinking about it, or, as in the old cliché, the skill of the centipede moving its legs normally but tripping if it thinks about it. Somatic tacit knowledge can enable a match official to appraise a rapidly unfolding situation in an instant without a conscious calculation. But with TV replays, this skill becomes redundant as time is effectively slowed, leaving viewers with experience of a sport, and who are well versed in the rules, at no disadvantage with respect to the major components of epistemological privilege. Since umpires and referees do make mistakes, and sometimes make glaringly wrong decisions, TV turns what would once have been presumptive justice into transparent injustice.

For a long time it was accepted that home viewers (and more recently spectators at the live game where a large television screen offers the broadcast replays) might have a better ability to judge than the umpire, but nothing was taken to follow from this: the location and process of decision making was not affected. This could be, and was, justified in the discourse of sports viewers and commentators by accepting that the umpire or referee could be "unsighted"—his or her view obscured by other players—and/or that the job of making decisions in real time was a hard one. It was possible to accept that the location of ontological authority should remain with the umpire and made compatible with evident mistakes resulting from this shift in epistemological privilege, so long as the mistakes were easy to understand given the difficulty of the umpire's job. As the recognized location of ontological authority, the match official's role is sometimes *defined* as *making* reality in the course of play irrespective of what TV shows, and thus match officials are banned from looking at replays. Sometimes replays are not shown on the ground but only on broadcast television, giving

the home spectator and TV commentators more epistemological privilege than anyone else; meanwhile, the game proceeds
as usual, because no one on the field knows what the television
is showing—surely what can only be a short-term solution, like
putting a thumb in the hole in a leaking dam.

We now need to step back and take a wider look at the technological aids to decision making in sports.

2   Sports Technologies Classified by Their Complexity

Measurement technology in sports is a matter of capturing events. Match officials have to deal with fleeting moments; we want to grab the event as it passes and hold onto it for further examination so as to further enhance epistemological privilege. We can call sports measurement technologies "capture devices." The moment of capture can also be time-stamped for comparison with the times of other captured events, such as in the case of the start and end of a race. We will now examine the technologies used in different sports and classify them in terms of five degrees of indirectness, as shown in table 2.1. We are going to use yet another new term and talk about "intermediation" instead of "indirectness," because "indirectness" is a bit vague and all-inclusive. What we really want to get at is something more positive—the amount of "stuff" or number of processes that stand between the event itself and the view we get of it once it has been grabbed and held for further examination on a capture device. We want to look at how much is going on in the intermediate space between the event itself and the captured event that we get to look at in a more relaxed manner. This "how much is going on" is the "degree of intermediation."

Table 2.1
Classification of capture devices

Level of intermediation	1	2	3	4	5
Examples	Bails, golf-cup, high-jump bar	Photo finish	TV replay	Manipulated TV replay, Hot Spot, Snicko	Track estimators

Level 1 technologies are the most direct: they have the lowest level of intermediation between the event itself and its frozen counterpart. Intermediation increases steadily as we go from Level 1 to Level 5. With more indirect technologies, the captured event is more and more distant from the original; more things have to happen in order for the event to be captured.

Level 1: Minimal Intermediation

The cup on the golf green is a good example of a Level 1 capture device: there is almost no technological intermediation between the ball reaching the target and its fall into the cup where its position is held for all to see. Other Level 1 devices include the high-jump bar, the sand in the long-jump pit, the embedded javelin, the mark in the turf caused by the discus or shot, the surrounding wall in baseball, the chain between two poles in American football, and the pockets on snooker and pool tables. In other sports, the signal created by the capture device may not last as long, but the principle is the same. Thus, even though a basketball net does not trap the action, it does slow it down enough to capture it for all practical purposes; something similar applies to the goal net in association football and the winner's tape in running events. In these cases, the technology enhances the epistemological privilege and ontological authority of match officials.

Some capture devices *redefine* the event to make it easier to trap. Consider cricket. We will return to cricket again and again as we work our way through the book, explaining it in greater and greater detail since it is a vital example of the use of sports technologies.[1] We will start our explanation of cricket by looking at the differences and similarities between it and baseball.[2]

What we will refer to as "the basic transaction" in cricket is very similar to that in baseball, in that it involves someone projecting a ball and someone trying to hit it with a bat. In cricket, the person projecting the ball is called the "bowler" rather than the "pitcher," and what the bowler tries to do is get the batsman or batswoman out. The batsman or batswoman is the equivalent of the batter in baseball. Because "batsman or batswoman" is clumsy, we will adopt the baseball convention and refer from now on to the "batter" even when it is cricket and not baseball we are discussing. Cricket is, however, a purists' game, and we apologize to the purists for our abuse of terminology.

One way of getting the batter out in cricket is for the bowler to hit the "wicket" with the ball—there is no equivalent in baseball. The wicket (figure 2.1) comprises three vertical stumps—lengths of wood with a circular cross-section about an inch-and-a-half in diameter and twenty-eight inches tall, driven into the turf side by side so as to make a target nine inches wide. Thus the batter tries to defend the stumps, known collectively as the wicket. On some occasions it will be difficult to tell whether the ball struck the wicket with a grazing or gentle blow or missed by an imperceptible margin. To settle this kind of issue, two small sticks called "bails" sit in shallow grooves on top of the three stumps;

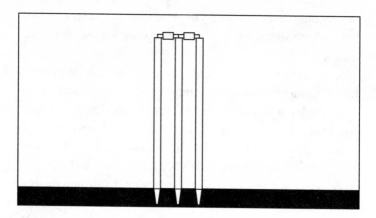

Figure 2.1
The stumps or wicket (not to scale).

the notion of striking the wicket is redefined as "breaking the wicket," which means one or both of the bails must fall to the ground. Thus the falling of the bails traps the event—either the bails are both still balanced or at least one of them is on the ground—while redefining the event as a matter of dislodging the bails.[3]

But life is always complicated. Even with the use of the simplest Level 1 technology, things can go wrong. The bails may be affected by a gust of wind just as the ball passes, dislodging them even though the ball just missed the wicket. The golf ball sometimes teeters on the edge of the cup, and the player is not allowed to wait too long for it to fall but must walk up to it and tap it in without delaying overlong; it is the job of match officials to interpret what is to count as an "acceptable" delay. In football, the net does not always bulge when a goal is scored, as the ball can cross the line and bounce back out without touching it.[4] It is also true that a high-jump bar can bounce around and fall off some time after the athlete has landed, and we need ways to judge what to do. In other words—and this is an important point to which we will return repeatedly in this chapter—there are cases where even the simplest technologies have to give way to, or be supplemented by, human judgment. Traditionally, in these cases, match officials have called on their ontological authority to determine the outcome.

Level 2: Minor Intermediation

Now we come to capture devices that involve a little more intermediation. The photo finish in horse racing is an example of a Level 2 technology: the positions of the horses at different times are presented in a single image captured by a special camera looking along the finish line. After some developments to the technology—going from a horizontal shutter to a vertical strip opening, from a still frame to a panning device and illuminating the finishing tape on dull days—the photo finish seems to have been readily accepted. The photograph is indirect because the horses are separated from the film on which their image is captured by the passage of light—unlike the golf ball, they do not fall into a pit when they reach their goal. Nevertheless, the technology is well understood and widely trusted. For example, no one worries that light takes a little longer to travel from the far horse than the near horse so the far horse might be around 1/10,000 of a millimeter, give or take or zero, further ahead than the photo finish shows. Compare this with the start of sprint races: the delay in transmission of the sound of the sprint-race starting pistol could disadvantage the far runner by hundredths of a second—and sprints are nowadays timed to hundredths or even thousandths—so in important races, each lane has its own loudspeaker with the signal transmitted across the track at the speed of light (electricity).

Crucially, however, the image produced for the photo finish still has to be examined and interpreted by a match official in order to determine who has won. In many cases, this is relatively straightforward—as it should be—but there are cases where the judgment and interpretation of match officials is crucial in explaining and justifying the decision. For example, in the 2014 Winter Olympics at Sochi, a close photo finish in which three skiers appeared to cross the line simultaneously was eventually awarded to Russia's Egor Korotkov because he threw his arms in front of him and crossed the line head first rather than feet first.[5]

Turning to the spectators, from their point of view, capture devices help them share in the decision. They can see the golf ball topple, the bails fall, and the net bulge. The photo-finish image is a public document that explains why the match officials classified the result one way rather than another. The important point here is that the technology is trusted to gather the necessary data in a sufficiently accurate and reliable manner.

Level 3: Moderate Intermediation

Matters become more complicated when several different pieces of information have to be integrated. Returning to cricket, consider the "run out." This is akin to the decision in baseball as to whether a runner has made his or her base before being tagged. (See figure 2.2 for a diagram of the field of play in cricket.)

A crucial difference, however, is that in cricket the batter does not score by running around bases on a diamond but by running directly toward the place from which the bowler projected the ball. Another set of stumps have been driven into the ground at

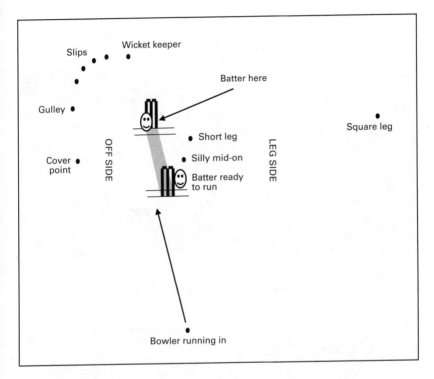

Figure 2.2

A cricket field with a very attacking field—all fielders are very close in. The field is set for a right-handed batter. For a left-handed batter, the offside and leg side (see chapter 3) and the fielding positions would be a mirror image. The boundaries are around eighty yards away; normally the field would be more spread with quite a few patrolling the boundaries and with a defensive field there would be fewer slips and no short leg or silly mid-on. The creases are shown with the gray box being the virtual space between the wickets. See figure 2.3 (and the figures in the "Bonus Extra" chapter) to see how this looks in real life, or search YouTube for more examples.

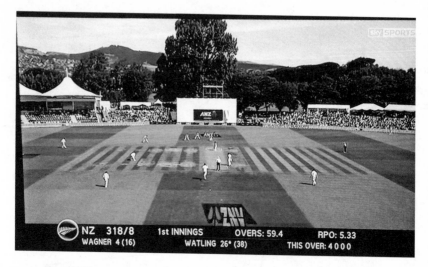

Figure 2.3
The field of play in cricket (Australia vs. New Zealand, February 2015).

Figure 2.4
A fast bowler delivers the ball.

that point, twenty-two yards away. The batter must reach that ground before that wicket is broken by the ball to score one run (runners are not tagged in cricket; rather, to get the batter out, someone must break the wicket with the ball before the runner makes his or her ground). If the batter has time to return and regain his or her ground at the end from which he or she started before *that* wicket is broken with the ball, then he or she scores two runs. If he or she can turn again and reach the far end once more, he or she scores three runs. On rare occasions he or she might return yet again to score four runs. A further complication in cricket is that two batters are on the field at once, one facing the bowler and one relaxing at the bowler's end, awaiting his or her turn while the bowler bowls (unlike baseball, double plays are not allowed). To complete a run, *both* batters must make their ground at the opposite end from where they started before the corresponding wicket is broken; a strong hit will see both run-ners running backward and forward, passing each other in the middle of the "pitch"—the strip of ground that separates the wickets. If an odd number of runs is scored, the other batter will usually take over the batting, because the batters will now be at opposite ends. But bowlers change over every six balls bowled with the new bowler coming in from the other end, and the positions of the fielders switches around with him or her—six balls is known as an "over"—so it won't always be the case that a new batter faces the bowler if an odd number of runs is scored; if the runs were scored at the end of an over, then the old bat-ter will be at the facing end. In yet a further complication, the batters do not have to quite reach the far wicket to complete a run, but rather need to reach a line drawn across the pitch four feet in front of the wicket known as "the crease." To avoid being "run out," the batter—some part of the batter-plus-bat—must touch the turf beyond this line prior to, or simultaneous with, the wicket being broken. The wicket can be broken by the ball or by a player's hand, so long as the ball is held in that hand. It doesn't happen very often compared to the other ways of being out, but the run out does happen once or twice during a game.

The "stumping" method of being out is closely related to the run out. Behind the batter stands or squats the wicketkeeper, roughly in the same position as the catcher in baseball though sometimes much farther back. The wicketkeeper's job is to the catch the ball if the batter fails to hit it and it misses the wicket. Sometimes an ambitious batter will "leave the crease" and take a few quick steps forward when the bowler bowls so as to be in a good position to hit the ball really hard. But if the batter misses the ball, the wicketkeeper can grab it and break the wicket, catching the batter out of his or her ground. It is just like the run out except that the batter wasn't trying to get a run, just get in position to strike the ball really well. What often happens in these circumstances is that the batter misses the ball, then turns and makes a desperate dive with bat outstretched to try to regain the crease before the wicket is broken; it is then up to the umpire to decide whether the wicketkeeper or the diving batter has been successful. Full-length diving with an outstretched bat is also often associated with attempts to avoid a run out, but in these cases the batter is running at full speed before the dive so it is even more spectacular and bruising.

The point to note is that in the cases of run outs or stump-ings, it is not only a question of judging the *exact moment* when the wicket is broken—itself a much more difficult thing to judge than whether it has been broken at all—but also whether it was fairly broken (e.g., it would not be fairly broken if the wicket-keeper or fielder broke the wicket while the ball was spilling out of his or her hands) and the exact moment a third event took place four feet away—the batter's grounded-crossing of the crease. *In most cricket matches*, there is no way to capture the exact moment the wicket was broken, the fairness of the wicket-breaking, or the split second at which the grounded-crossing of the line occurred. Instead, the ontological authority of match officials rests solely on their epistemological privilege: using their own perceptual abilities—somatic tacit knowledge—and viewpoint on the pitch, they must decide and categorize in real time and without the benefit of any technological support; of course, they have good epistemological privilege.

In *televised games*, however, technology can be used to assist the umpire: TV replays can provide a Level 3 capture device. It is a little more indirect than a photo finish because of the electronics in a TV camera, and a little more fallible than the photo finish because the technology was not designed for the sole purpose of assisting the umpires. As a result, the position of the cameras is less than optimal, the frame rate is slower than would be preferred for a purpose-built device, and more manipulation of the images—for example, zooming in and out, replaying in slow motion—is needed to provide the necessary information. In a typical stumping or run-out decision, a dedicated match official, known as the "third umpire," has access to the various broadcast camera feeds; by zooming in, or selecting different views, he or she reaches a judgment about whether the batter should be "given out." This decision is then relayed to the on-field umpire and to the spectators. As with the photo finish, the TV footage used to reach the decision can be made available to spectators, and, again like the photo finish, the technologies involved are pretty well understood and widely trusted. Needless to say, on some occasions the TV replays do not provide a clear image; but, when this happens, spectators can see the problem too and cannot reasonably feel that a great injustice has been done when the off-field umpire finally uses his or her best judgment to make the call.

Another kind of difficult judgment that cricket umpires often have to make is whether or not the ball just brushed the edge of the bat before being caught by the wicketkeeper. In cricket this counts as "a catch" and the batter is out. The difference between "caught behind" (by the wicketkeeper) and a foul tip in baseball, where the catcher catches a ball that has touched the bat, is nicely set out on Wikipedia (https://en.wikipedia.org/wiki/Foul_tip).

The foul tip is roughly equivalent to caught behind in cricket except that whereas in cricket a batter caught behind is immediately out, a caught foul tip only counts as one strike so a batter would only be out from a foul tip if he was already on two strikes. Caught foul tips are rarer than caught behind in cricket for two main reasons. The round shape of a baseball bat means that slight deflections are more likely to deviate significantly, making it more difficult to catch; by contrast, a cricket bat has flattish edges. Also, a baseball catcher must take position immediately behind the batter, meaning that he has less time to react to a tip. There are no restrictions on where the cricket wicketkeeper stands, and he or she can often be as much as fifteen yards behind the batter, giving him or her more time to react to edges, especially when facing fast bowlers. Furthermore, it is not unusual for there to be extra fielders beside the wicketkeeper, called "slips" (see figure 2.2), who can catch bigger deflections. Extra fielders behind home plate are not permitted in baseball and would probably be of little use anyway.

It should be borne in mind that in cricket the area of the field of play behind the batter is every bit as much part of the playing area as the area in front or to the side: there is no "foul area." As a result, one elegant shot that a skillful batter can play is the "late cut," where the ball is deflected very slightly as it speeds past the batter—the less deflection, the harder the shot—so the motion imparted by the bowler carries it past the wicketkeeper into the part of the field behind the batter, thus allowing runs to be scored. The "leg glance" achieves the same effect but on the other side. Other batting strokes, the meaning of which we discuss in the "Bonus Extra" chapter toward the end of the book, include the cut, the square cut, the hook, the straight drive, the on-drive, the lofted drive, the sweep, the slog-sweep, the slog to cow corner, and, introduced more recently in response to the emergence of the limited-overs game (see again the "Bonus Extra" chapter), the reverse sweep and the ramp shot (but don't worry about them).

But perhaps the ball that seemed to touch the edge of the bat with hardly any deflection, leading to an apparent caught behind, actually just *missed* the edge! The visible effects of such slight contact are very hard to see in real time and are often below the resolution and frame rate of the TV cameras. As a result, television replays can provide little extra information and rarely provide a sound basis for overturning the on-field umpire's decision. Level 4 devices (see below) have been introduced to try to resolve this kind of problem, but, in their absence, spectators can at least see that, whatever the umpire's decision, he or she has made no glaring error—given the resources available, the umpire could not have done any better. There is no obvious injustice, and ontological authority resolves the issue.

More seriously, in some circumstances TV replays are known to be misleading. One is the decision over whether a ball has been fairly caught. As in baseball, a player is out in cricket if the ball is struck and then caught by a fielder before it bounces. In cricket, by the way, the fielders, other than the wicketkeeper, do not wear any kind of gloves—it is all bare hands. With a low fast catch it is sometimes not clear if the ball flew straight into the fielder's hands or hit the turf just in front of the hands before being caught. All commentators agree that the foreshortening effect of TV replays means they often appear to show that the ball did bounce, and many commentators point out that this kind of TV replay almost always gives the benefit of the doubt to the batter, who is then given "not out."

Level 4: Significant Intermediation

Level 4 devices differ from Level 3 devices in that the intermediating technologies do more than re-present events that were, in principle, available to an on-field match official with perfect vision and discrimination. Instead, Level 4 technologies generate new data that would enhance even the most perfect human's perceptual abilities by allowing him or her to see things from new angles and via nonvisible wavelengths and transpose sound into a visible waveform. Level 4 devices can resolve the "caught behind" problem in cricket.

In cricket, Hot Spot is a Level 4 device that uses infrared cameras to detect the heat produced by contact between bat and ball and indicates whether the ball has touched the bat in case of a disputed catch. Impact between bat and ball will show up as a bright spot on the edge of the bat in the black-and-white infrared image captured by a special TV camera. Here the intervening medium is infrared light, which makes this technology more complicated and less well understood than standard photography. Higher-level capture devices, because they are complicated and involve novel applications of new technologies, tend to fail in unexpected ways. Thus, only after some years was it realized that Hot Spot may not work well if the weather is very hot and that it may fail to detect a slight contact if the edge of the bat is greasy.

As time has gone on, Hot Spot is often complemented by another Level 4 technology—"Snickometer" or "Snicko," or some other propriety version such as "UltraEdge"—that generates an oscilloscope-type image of the sounds made at the crucial moment. This chart is synchronized with and superimposed over TV film to show whether the sounds heard were caused by contact between ball and bat or some other object. Because the shape of the trace will be different for different kinds of contact, Snicko seems to enable umpires to distinguish between the ball hitting the bat—a sharp, spiky trace—and some other kind of contact, such as the ball striking the batter's clothing or pads (see below), the bat striking the ground, and so on, which will produce a more "woolly" trace. When used together with TV replays, Hot Spot and Snicko (see figures 2.5 and 2.6) seem to provide an acceptably reliable combination of Level 3 and Level 4 capture devices and to perform a useful function in correcting errors and reinforcing the epistemological privilege and ontological authority of match officials with the process viewed and understood by the TV audience.

Pundits and commentators also use a combination of capture devices to reveal the accuracy of offside decisions in football. Here the simultaneity of two events has to be judged, but the events are not colocated and may be as far apart as the length of

Figure 2.5
Hot Spot.

Figure 2.6
Snicko.

the football field. The two events are the moment the ball leaves the foot of the player passing it forward and the position of the receiving player in relationship to other players *at that moment*. At that moment, no part of the receiving player's body must be farther forward than the last opposition player (goalkeeper aside). The optimum position from which to judge the relative position of the receiving player is level with that player, while the optimum position from which to judge the moment the ball was kicked is level with the kicker. The duty of making this judgment is given to a specialist "assistant referee" (once called a "linesman") who is advised to patrol the sideline, keeping up with the front players. TV replays that capture the whole field can usually reveal the moment when the ball was kicked, but they are less good at indicating the relative position of the front player, because the view of the TV camera will almost always be at an angle rather than along the line of the front players.

Introducing more intermediation moves us nearer to a solution to the offside problem. Computer graphics techniques can be used to capture the crucial TV frame as a virtual image, rotate the scene electronically, and draw a virtual line across the field that is level with the front players at the moment the ball was kicked; the TV viewer seems to be able to look along this line. BBC's *Match of the Day* program uses such transpositions, but they do not, as yet, influence the decisions of match officials. There are no published statistics about the accuracy of such methods and the BBC would not respond to our requests to discuss this with them, so we cannot be sure how accurate they are—but we have no reason to doubt their accuracy. One might argue that the accuracy of this device does not matter, as it is not used by referees. But it does appear to reduce the epistemological privilege of referees and erode the grounds of their ontological authority, so it is important nevertheless, and we should know how well they work. We will assume they work accurately, but we do not know.

Level 5: Maximum Intermediation—Track Estimators

Now we move to highest level of intermediation, in which technologies—track estimators—are used not just to record or represent data but to generate simulations of the on-field events. Hawk-Eye was the first and probably the most well-known of these technologies. Using specialist cameras, track estimators gather data that indicate the position of the ball in play and try to reconstruct its track in computer-generated graphical form within a virtual version of the playing area complete with boundary and other play-defining lines. The resulting animated computer graphic is taken to show the path of the real ball and, crucially, exactly where it bounced, or struck some other feature of importance such as a batter's leg. We will discuss the use of track estimators at length in chapters 3 and 4. With the appropriate use of computer-generated images, the reconstruction offered by a track estimator can appear to be the equivalent of a television replay, but a much greater amount of inference, calculation, and computation stands between the events and reconstruction than stands between TV cameras and broadcast video. Also, the position of the reconstructed ball within the simulation is always judged in relation to the virtual model of the playing area and not against the actual playing surface, while the bounce footprint of the ball is a mathematical model, not a recording of actuality.

We will draw on this classification of technological aids in terms of levels of intermediation throughout the book. What we

will try to establish is that, other thing being equal, less intermediation is better than more intermediation because the higher
up the scale of complexity you go, the more expensive things
are, the more things there are to go wrong, the more the public
can misunderstand what is going on, and the less the technically
assisted game is like the game the rest of us play. We will argue
that more complex solutions arise out of an obsession with accuracy rather than justice; to resurrect justice in sports, we need
to return as much ontological authority to the match official as
possible.

3 Track Estimators and Cricket

Humans and machines make errors of two kinds. "Systematic errors" are repeated errors that have a similar effect each time; we can often understand the causes of these errors and can predict and thus mitigate their impact. For example, suppose when I take my temperature I have a tendency to remove the thermometer from my mouth a little too soon: in that case, the temperature I think I am measuring will always be a little low. But if I discover that it is always a little low, I can work out a way to fix the problem—remind myself to keep the thermometer in my mouth for a bit longer, say. "Random errors" are more difficult to deal with. We cannot predict them, although we can estimate their typical size and the way they are "distributed"—as will be explained below.

We now want to apply this idea of error to track estimators such as Hawk-Eye. The problem the public faces is that track estimators are often presented as though they simply show what happened, without any error at all. But all machines make errors, just as all humans make errors. Engineers and scientists find no shame in this. Among scientists and engineers, it would be thought strange if a measurement were presented without an estimate of the error associated with it. This is just the scientific equivalent of you taking your temperature and seeing that it is 98° but knowing that the real temperature could be anywhere between 97° and 99°, because your thermometer is not that accurate and it is never quite clear if you have put it in the right place for the right time, and so on. Scientists and engineers have much more careful ways of dealing with these kinds of things.

What we need to do is make our own estimates of the kind of errors we think track estimators make, because not enough information about these errors is available to the public. It is true that some remarks about error can be found on Hawk-Eye's website, even though in a sport like tennis the results are presented in real time as though they were perfect; but we are going to try to explain this question of error a little more in the way that scientists and engineers think about it, and then take another look at the way the track estimators should be showing their results.

We'll start with the idea of an "error distribution." The idea is simple—if you take your temperature a large number of times you will hardly ever make a very large error, like getting the temperature wrong by two degrees, but you will make lots of small errors—a tenth of a degree here, a half a degree there. The idea of an error distribution is that small errors happen much more often than large errors, and you can show this on a diagram. If we create a diagram that has the number of errors represented by height on the vertical axis and the size of the error represented by distance from the true value on the horizontal axis, we will get what is called a bell-shaped curve—the shape of a bell if you cut it in half. Lots of kinds of measurements have bell-shaped error distributions; the most common shape for the bell is what is called a "normal curve" or "normal distribution." A normal distribution is shown in figure 3.1.

You can imagine this curve represents you measuring your temperature thousands of times. You would get it right a lot of

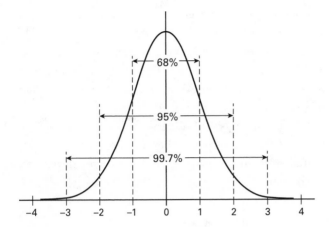

Figure 3.1
The normal distribution.

the time—the peak in the middle of the curve—and you would get it wildly wrong only very rarely, sometimes much too high—the tail at the right-hand side—and sometimes much too low—the tail at the left-hand side. The rest of the time your results would be in between the extremes, with many measurements close to the center and fewer and fewer as you move out and the level of error increases. The other lines in figure 3.1 show that in a normal curve 95 percent of all the measurements will be in a range bounded by two vertical lines set at a certain distance from the center. This distance is "two standard deviations," and the other lines and numbers show how many of your measurements would be within one standard deviation, three standard deviations, and so on. But we do not need to worry about standard deviations; we need only to grasp three ideas:

(1) Even if errors are not distributed exactly normally, they will nearly always follow the pattern of this rough bell shape, and that means the bigger the error the fewer there will be.

(2) Let us say we know the average size of the error: imagine I took all my thousands of temperature measurements and I could somehow find out how far each one was from my true temperature and then I worked out the average of all the errors I made. I would know for sure that some of my temperature errors would be smaller, and, crucially, some would be larger. Some—just a few—would be a lot larger. This is very important because track estimators like Hawk-Eye, when they do discuss their errors, present an *average* error, never mentioning that the device can be expected to make larger errors from time to time.

(3) When a measurement is presented it should always be presented with some estimate of possible error.

Errors arise out of a whole raft of different causes. In the case of temperature, I might have just drunk a hot cup of tea so the temperature would read too high; I might have eaten ice cream, making it too low; I might have breathed a lot, cooling the thermometer while it was in my mouth; I might be too hot because I just ran five miles; I might have put the thermometer above my tongue or below my tongue or under my armpit; and so on. And, of course, lots of things might contribute to the final reading that we cannot even guess about. That is the way in which the bell shape can arise—big errors emerge (rarely) when lots of things combine, and small errors occur (more often) when only one factor is present. What we need to do is work out all the possible sources of the track estimators' errors and, also, how they might add up.

The Technology of Hawk-Eye and Other Track Estimators

As we started to explain in chapter 2, Hawk-Eye, when used to adjudicate calls in cricket and tennis, is a video-processing system that uses a number of cameras and a computer to store and process the data. The original patent application indicates that six cameras are used but also says that only two of these are needed for the system to work; the others provide redundancy (e.g., if one camera is blocked by a player). These cameras track the flight of the ball, and the computer then uses their feeds to reconstruct the trajectory of the ball by analyzing the pixels in each frame of each relevant camera. The field of play is also modeled within the system, as are some of the rules relating to the game. By combining the trajectory of the ball with the model of the pitch and the database of rules, the path of the ball can be reconstructed against the background of the main features of the playing area as a virtual reality, and a decision can be given (e.g., should the batter be given out in the case of cricket or, in the case of tennis, did the ball land inside or outside the line). The reconstruction can be shown to television viewers.

The cameras will have a limited frame rate, and the position of the ball between frames has to be reconstructed by the software. In the patent application, this process is not described in detail, but it is clear that the analysis is statistical, with predicted and measured paths being compared in order to generate the final trajectory.

In our analysis, we follow Hawk-Eye's practice in describing how it works: its website presents the device as calculating a trajectory from a series of images of the ball captured in separate television frames.[1] We are unconcerned with the exact statistical algorithm used in the reconstruction of the path of the ball. We take it that, as in all science and engineering, whatever the algorithm, the accuracy of the final result has an upper limit set by the accuracy and amount of the data. The crucial data concerning the path of the ball, then, consist of a series of "data points." In the first instance, information is generated from camera frames containing certain pixels that are taken to signify the ball and other pixels that are taken to signify the line, wicket, or whatever is needed to locate the ball-pixels in the space of the playing area. When we refer to a "data point," we mean a three-dimensional reconstruction of the position of the ball that involves information from two or more cameras. The idea of a data point serves the analysis adequately whether or not such data points are ever actually constructed as the Hawk-Eye processor runs. When we refer to "frame rate," we are talking about individual cameras. It might be that the "effective frame rate" of Hawk-Eye is higher than that of a single camera because it combines frames from more than one camera to establish the trajectory of the ball in any one dimension. We simply do not know. Therefore, we assume that the "effective frame rate" is whatever frame rate we find mentioned in our sources. Any rough calculations we make are based on these seemingly plausible assumptions, but they are open to correction.

It would improve this analysis to have the answers to certain questions, which we will lay out in numbered form as we go through this chapter and the next. The first two questions bear on frame rate:

Question 1: What is the frame rate of the cameras used by Hawk-Eye and other track estimators, and how has this changed over the history of the devices?

Question 2: Does Hawk-Eye (or other track estimators) use feeds from regular cameras or does it use dedicated cameras, and how has this changed over time?

In our language, then, each effective frame provides a single data point; reconstructions will be more accurate with higher frame rates, as this minimizes the distance between data points. In addition, having more data points should allow more complex curves to be detected . Thus, two data points could provide information that can be used to infer the straight-line direction and velocity of the ball (subject to errors). For curved trajectories, more data points would be needed; and for trajectories where the rate of curvature changes, still more data points would be needed. It follows that, given a certain frame rate, the faster and curvier and more rapidly changing the flight of the ball, the more likely the reconstruction of its path will be subject to error.

In some circumstances, Hawk-Eye projects a hypothetical trajectory beyond its last data point. One such case is the "leg before wicket" (lbw) decision in cricket in which the continued trajectory of the ball beyond the last data point is projected forward and then represented on a virtual reality display that shows the television viewer what "would" have happened had the ball not hit the batter. We start with the lbw situation because the technicalities (if not the game) are relatively easy to understand, and it was the lbw decision that gave rise to our initial puzzlement about the claimed accuracy of track estimators. But before we proceed, let us explain our methods and why we have all these unanswered questions scattered throughout the chapter.

Why All These Questions?

This chapter and the next two are about track estimators. Here we concentrate on Hawk-Eye as it was the first in use. Hawk-Eye was first introduced as an aid to judging the lbw rule in cricket, and this is why cricket is so central to this book. Track estimators are also well known for the use in tennis, and here we will compare the use in the two sports. Hawk-Eye, because it has been used for several years, has a well-developed website and has been the subject a range of media coverage and a few published articles. In what follows, nearly all the material we use in the analysis is drawn from such sources and so can be readily checked. Aside from some initial inquiries, we were unable to obtain significant information directly from Hawk-Eye Innovations, or from the makers of Hot Spot or the BBC in respect of the *Match of the Day* reconstructions of offside (see below). We did get considerable help from the makers of Tennis Electronic Lines and Auto-Ref, which are other technological devices applied to match officiating. We discovered toward the end of the analysis that a number of our questions and proposals had already been put by contributors to newspaper websites, but we cannot find any detailed response to the newspaper queries either, suggesting that our experience of meeting a near blank wall in the face of technical inquiries is not unrepresentative. (More details of what we were and were not able to find out about the Hawk-Eye system can be found in appendixes 2 and 3.)

The main professional work of the first two authors of this book is the analysis of science and technology. We have always found academic science readily accessible, and we have found academics ready to answer even our most intrusive technical questions. Thus Collins has spent around forty-five years studying the science of the detection of gravitational waves, writing four books about it, and has almost never experienced any problems of access to material. The secrecy we encountered when we set out to investigate track estimators is quite untypical of two lifetimes of experience working at the boundaries of science and technology. We consider that the secretiveness of the sports-measurement enterprises is dangerous and not in the public's interest; the defense we have most often encountered when we have asked for details of the statistics of tests carried out is "You can't have them because of commercial propriety." This can only invoke deep suspicion.

Suspicion is not science, however, and to gain as much information as we could, we used the major search engines and databases systematically to search popular websites and the academic literature. Specifically, three people spent a total of about fifteen hours searching Google and Google Scholar (we looked at the first twenty pages that were returned) plus Web of Science (there were only three, irrelevant, hits) for articles relating to Hawk-Eye (spelled in various ways).[2] Of the information we uncovered, we found the most useful on newspaper websites, a discussion site called "Cricinfo," and Hawk-Eye Innovations' own website. We also examined the original patent application, which is available online from the European Patent Office (reference number WO 01/41884). We also discovered an article in *Scientific American* (Fischetti 2007) and a published analysis of line calls in tennis (Mather 2008).

We found only one article about Hawk-Eye in an engineering journal but it was very hard to access, in both senses of the term. We discuss this paper more fully in appendix 3. We do not know how that paper, which reports on the development of a device under contract with Hawk-Eye Innovations, bears on the technology that is currently used, but, in any case, no substantive difference would be made to the analysis if it included more reference to this paper unless it is true, as the paper implies, that Hawk-Eye takes its television feeds from existing TV network cameras; this would indicate lower camera frame rates and a greater potential for error than we discuss in this chapter. In any case, we are sure the technology has moved on.

Reliance on data in the public domain means that our information about how Hawk-Eye works is not complete. If we had the kind of access that we have to the academic scientists and technologists with whom we normally work, we would have expected to gain answers to a number of questions; but now those questions are distributed throughout these chapters and, unfortunately, have not been answered. As we work through the analysis, we show where the answers to these questions would be useful. As it has turned out, then, we have mostly had to rely on "back-of-the-envelope" calculations based on what we can glean from the public domain, but, as will be seen, quite a lot can be learned using only this approach.

The Leg before Wicket Rule and Track Estimators

Turning back to the substance of our analysis, we now have to explain a bit more about cricket. To start with, bowling in cricket is quite different from pitching in baseball, for many reasons. For one, the bowler nearly always aims to hit the ground with the ball before it bounces up into the domain of the batter. Therefore, a crucial feature of bowling is the way the ball bounces off the pitch. Slow bowlers will aim to impart a lot of spin to the ball so that it jags sideways when it hits the ground, and faster bowlers will cause the ball to swing through the air both before and after it hits the pitch and also jag sideways off the pitch as a result of interaction between the pitch and the raised seam of the ball.

Another crucial difference between cricket and baseball is that in cricket there are no "balls" or "strikes." There are illegitimate balls, as when the bowler oversteps the mark or bowls the ball beyond the batter's reach, but these are rare exceptions rather than a regular part of play. Most important, in cricket matches there is no limit on the number of balls a batter may face; as far as the batter is concerned, the game lasts for as long as he or she can avoid being out. In longer forms of the game, batters have gone on for days; and even in the more frenetic short forms, the batter is always balancing the need to score runs quickly with the need not to be out. But there is nothing in the rules to make the batter end his "innings" if he does not score any runs.

As we have seen, if a ball from a bowler hits the wicket, the batter is out. The batter wears pads to protect each leg. Each pad is an armored sheath running from ankle to just above the knee. Allowing the ball to hit the pads is an integral part of the game; the batter would never be out if he or she simply stood in front of the wicket, blocking the path of the ball, kept the bat out of the way, and allowed the ball to hit the body or the pads every time. The notoriously complicated lbw rule stands in the way of this possibility. It says that the batter is out in certain restricted circumstances if the pads alone stop a ball that would otherwise have hit the wicket—this counts as out in virtue of leg before wicket or lbw.

We now need to go deeper into the heart of cricket's darkness to explain the restricted circumstances that apply to the lbw rule and upon which the use of track estimators in cricket turns. Cricket, as the old hands say, is a sideways game. The bowler will deliver the ball from a sideways position like a javelin thrower and the batter will stand sideways to receive it (see figure 2.2 for a diagram of the field of play in cricket). A right-handed batter's left leg will face up the pitch toward the bowler, with the right leg nearer to the wicket. A left-handed batter will be the other way round. Though batters are not exactly sideways, this roughly defines their orientation, and this orientation is important because it defines the sides of the pitch in relationship to the stance of the batter. The side of the pitch closest to the batter's hands and bat is called the "offside," whereas the side closest to his or her legs—the side on which he or she is standing—is called the "leg side" or "onside." The "late cut," which was discussed in the last chapter, is a deflection toward the offside, whereas the "leg glance" is a similar deflection toward the leg side. To understand the lbw restrictions one must imagine a rectangular strip, nine inches wide, painted on the wicket and stretching between both wickets (or drawn on the "pitch," if "wicket" is too confusing). First, a batter cannot be out in virtue of lbw, even if the ball was going on to hit the wicket, if the ball pitches on the leg side of that strip, unless he or she was not trying to hit the ball with the bat (playing a stroke). Second, a batter cannot be out by lbw, even if the ball was going on to hit the wicket, unless the ball strikes the batter's pad "between wicket and wicket"—that is, in the area above that imagined rectangular strip. Bear in mind that the ball can jag sideways off the pitch and the bowler can bowl from quite a wide position so that a ball that is going on to hit the wicket could hit the batters' pads well outside the space above our imagine painted rectangle. Once more this applies only if the batter was "playing a stroke"; if just standing there with bat raised or hidden behind the pads, then the batter can be out "leg before wicket" even if struck outside of the wicket-to-wicket area above the strip. There is room for a lot of interpretation about whether a batter was playing a stroke—after all, it is all about the batter's intention.

So, those are the restrictions. If those restrictions are satisfied, or if, in the umpire's judgment, the batter was not playing a stroke, the batter is out if the ball hits the pad without hitting the bat first and was then going on to hit the wicket. For all nontelevised cricket, an umpire, who stands at the point from which the bowler bowls the ball, is the sole judge of whether the ball (a) falls within the restrictions and (b) would have gone on to hit the wicket. Nowadays, in top games, a track estimator can be called in to help the umpire's decision. The track estimator will create a reconstructed image of the playing area with a virtual rectangle drawn between wicket and wicket and then reconstruct the track of the ball and project it onto the playing area. The reconstructed graphic shows where the ball is estimated to have pitched and, in particular, whether it pitched on the leg side outside of the virtual rectangle—ruling out lbw so long as the umpire judges that the batter was playing a stroke. (It can pitch outside the off side of the virtual rectangle so long as it hits the pads above the virtual rectangle and the batter was playing a stroke.) The track estimator also reconstructs the point at which the ball struck the batter's pad and determines whether that point was above the virtual rectangle—that is, between wicket and wicket. Still more contentiously, the track estimator reconstructs the path that the ball would have taken had it not hit the batter's pads, indicating whether it would have gone on to hit the wicket or not. Discussion of the accuracy of this kind of reconstruction, the kinds of error to which it is subject, and the way such errors are handled in cricket and tennis makes up this chapter and the next.

Figure 3.2 is a two-dimensional schematic version of the lbw situation from the side. The ball, traveling from left to right, bounces ("pitches") and then hits the batter's pad-protected leg. The dotted portion of the trajectory is what has to be judged or estimated. Television viewers see a three-dimensional virtual reality representation of the projected path of the ball against a virtual cricket field, and they can see where it pitched in

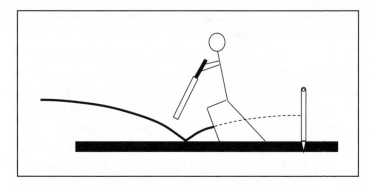

Figure 3.2
A two-dimensional schematic of a potential lbw situation.

relationship to the virtual rectangle drawn between wicket and wicket, where it struck the pad, and whether it was going on to hit or miss the wicket. For a number of years after the introduction of Hawk-Eye, cricket commentators would simply remark on what Hawk-Eye showed on the screen, giving the impression, perhaps inadvertently, that the virtual reality represented exactly what would actually have happened had the pad not been struck. This is where our analysis of Hawk-Eye begins.

The trajectory of the ball after it hits the ground can vary enormously. The bounce depends on the speed, hardness, and texture of the ball—which changes during the game—the state of the ground at the exact point of the pitch, the spin on the ball, and the position of the seam. The "swing"—the aerodynamically induced curve in the flight of the ball, which can be in any plane—depends on the ball's speed, its spin, its state, its orientation, the orientation of the seam, and the state of the atmosphere. As a result, what happens to the ball after it bounces is not predictable from its prebounce trajectory, so that the technology has to estimate the postbounce trajectory from

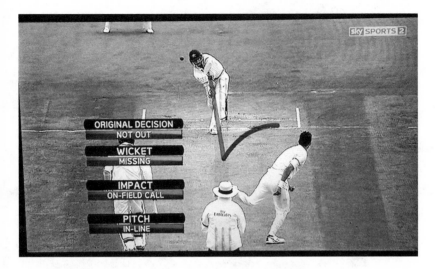

Figure 3.3
Track estimator in use for lbw decision.

postbounce behavior of the ball, for which it can gather data only between the bounce and impact on the pad. Paul Hawkins, then the director of Hawk-Eye Innovations, in response to a criticism from Dennis Lillee, the Australian fast bowler, explained the matter as follows:

"Hawk-Eye simply observes and then calculates the actual trajectory of the ball. Whether the cause of this trajectory was due to atmospheric conditions, the wicket, or the ball hitting the seam is irrelevant from a Hawk-Eye perspective. Hawk-Eye just tracks what happened—it does not try to predict nor to answer why it happened."

So, if the ball rears up unexpectedly after hitting the seam or a crack on the pitch, Hawk-Eye will track the trajectory off the pitch to predict the future course of the ball. Similarly, the tracking system will come into play if the ball shoots along the ground after hitting a dry spot on the pitch. (Rajesh 2003)[3]

Figure 3.4
Track estimator showing estimated path of ball from behind the wicket.

Our concern in analyzing what Hawk-Eye can do is to understand more fully what it means to "track" and "predict" the path of the ball. The accuracy of predictions is limited by, among other things, the quality of the data. To repeat what was established earlier in this chapter, no measurement is ever exact.

A decision is not a measurement. A decision is binary, like the "guilty/not guilty" decision of a jury; in cricket, the batter is either out or not out. The process of what we will call "digitization" is used to turn what are (always) inexact observations into discrete decisions. In most sports, the referees or umpires do the digitization, relying on their ontological authority; and what we are discussing here could be described as technical aids to this sort of digitization. The bails in cricket, for example, are one such aid to digitization. As discussed above, it can sometimes be difficult to tell whether the ball has touched the wicket and the falling of a bail converts this uncertainty into one of two discrete possibilities that, in most cases, have merely to be "read off" by the umpire.

In the case of lbw decisions, of course, the bails can't help, because the flight of the ball has been stopped before it gets to the wicket. Instead, the umpire has to judge whether the bails would have been dislodged had the ball continued on its trajectory and passed "through" the batter instead of being stopped by the pads. So, can a track estimator effectively replace the bails in the decision-making process?

This kind of technological digitization cannot correspond exactly with that of the bails in all cases, because in some circumstances it will be impossible to predict with certainty whether the ball really would break the wicket if it had not hit the batter's pad. To take the extreme circumstance where the ball just touches the wicket, whether the bails would fall would depend on how firmly they are sitting in their grooves, how rigidly the stumps are held in the ground (the tapered part at the base of each stump is hammered into the turf to hold them upright), or whether the bails and stumps are wet or dry. Again, the ball is not exactly spherical so whether it would cause a bail to fall can depend on its orientation as it passes the wicket. We assume that predicting all these things is beyond the capacity of both Hawk-Eye and human umpires; all we are doing in this paragraph is reestablishing *the principle* of the imperfection of all such measurements. It might help to have an answer to this question, however:

Question 3: Can Hawk-Eye or other track estimators measure and make use of the spin of the ball in cricket? Is it able to predict the orientation of the ball when it passes the wicket?

What we would like to know is what Hawk-Eye's and other track estimators' measurement errors look like. We would like to see a graph of the type that we have exemplified with human body-temperature measurement. We would like to know if the "bell" is narrow or wide and the size of the standard deviation. But we have not found any detailed indications of the dispersions of Hawk-Eye's errors in the public domain. The nearest we could find are the following quotations from Paul Hawkins when interviewed by S. Rajesh on the Cricinfo website:

Hawk-Eye has shown that balls pitched on roughly the same area on the wicket have passed the stumps at widely varying heights. And in tests conducted, thousands of deliveries were bowled from a bowling machine and filmed by Hawk-Eye. The camera feeds were cut about two metres from the stumps, approximately the point where the batsman would normally intercept the ball. When the ball hit the wicket, Hawk-Eye was able to determine, to within about 5 mm, the point of impact. ...

"Hawk-Eye requires between 1 to 2 feet of travel after the ball has pitched to be able to accurately track the ball out of the bounce (this is significantly less than an umpire requires). In instances when this does not happen, a Hawk-Eye replay is not offered to TV. (Rajesh 2003)

Watching a fair bit of cricket on TV, however, Collins has never encountered a case where Hawk-Eye did not provide a graphic when the bounce-to-pad distance was short; and so, another question we would like to ask is how often have such nonpresentations occurred? We would also like to know if a graphic has ever been offered even though the distance was very short and, in those cases, how the graphic was generated.

Question 4: How many times has Hawk-Eye or other track estimators been unable to provide an LBW decision or graphic because the distance between bounce and pad was too short to accurately predict the onward trajectory of the ball? What percentage of total Hawk-Eye judgments does this represent, and what is the figure for other track estimators? Are there occasions, and if so how many, when the distance was very short but graphics and decisions were generated? What percentage of total Hawk-Eye judgments does this represent, and what are the percentages for other track estimators?

Hawkins explained in 2006:

In most cases Hawk-Eye's output is accurate to within five millimetres in predicting the path of the ball. The accuracy levels are highest when the ball has travelled a fair distance after pitching, but even when the point of contact is very close to the pitch of the ball, the accuracy levels are still within 20 mm. (Rajesh 2006)[4]

Note that twenty millimeters is four-fifths of an inch, which is quite a long way when it comes to breaking or not breaking a wicket.

We don't know if the developers of Hawk-Eye have attempted to estimate the dispersion of their errors in a way that is more exact than is indicated in the above quotations, but we have not found any such report. This gives rise to Question 5:

Question 5: Has the distribution of Hawk-Eye's errors been measured? If yes, are the results available for consultation? If the errors have been measured, are they roughly normally distributed? If not, how are they distributed?

As the quotations intimate, the size of the error will be affected by the length of the bounce-to-pad trajectory, but it will also depend on other factors. The bounce-to-pad trajectory is not of fixed length; the position of the bounce is variable and deliberately varied by the bowler, and the batter can move forward or backward to try to reach a position from which he or she can most comfortably defend the wicket or strike the ball forcibly (hence opening up the possibility of stumping). The length of the projected pad-to-wicket trajectory depends, then, on the pitch of the ball and where the batter is standing when struck. The accuracy of Hawk-Eye's estimates is likely to depend directly on the distance between the bounce and the moment of impact on the pad—the longer the better, because then more data points can be gathered so as to give a better estimate of the projected path of the ball, especially where that path is not a straight line. There will be an inverse dependence on the distance between the impact on the pad and the wicket—the shorter the better, because the longer the distance the more damaging an error would be in the estimate of the path. To help with this problem we would like answers to Questions 4 and 5 (above) and also to Question 6:

Question 6: How, in addition to the 20 millimeter figure given for "short" bounce to pad trajectory, is the accuracy of Hawk-Eye and other track estimators affected by longer and shorter bounce-to-pad trajectories and pad-to-wicket trajectories? Is it the case that a human umpire could make a good decision that a track estimator would be unable to make in the case where a ball bounced very near to a batter's pad and the batter was well back and very near to the wicket?

In the case of the human umpire making an lbw decision, it is acknowledged that the accuracy of the judgment is affected by how close the batter is to the wicket when the pads are struck by the ball. If the batter whose pads are struck is well forward—a long way from the wicket—then, traditionally, an "out" decision is rare. In this way, *human judges deliberately introduce a systematic error into their judgments* that favors the batter—the so-called "benefit of the doubt" rule. The importance of this rule will become clear later.

As we have said, none of the information that we have found in the public domain gives any indication of the overall dispersion of Hawk-Eye's errors; the only indication of how the error increases with a short bounce-to-pad trajectory that we can find is given in the quotation above. We can find no systematic information about how the size of the error relates to the position at which the ball bounced, the speed of the ball, the length of pad-to-wicket trajectory, the length of bounce-to-pad trajectory, the degree of spin, the degree of swing, the nature of the pitch surface, and the nature of the atmosphere. Although the patent application does acknowledge that the distance of the camera to the ball is important (e.g., frame sizes are set to make the ball appear as large as possible) and that the position of the sun matters (e.g., it is important to distinguish the ball from its shadow), it does not tell us how variations in these or other parameters affect the accuracy with which the position of the ball can be tracked.

Unfortunately, the importance of accounting for these errors is greatest when the decision is most difficult. When the projected point of impact is well away from the edge of the wicket, there is unlikely to be any real doubt about the "correct" decision, and so Hawk-Eye's errors are probably considered "hardly worth reporting." In both cricket and tennis, Hawk-Eye's and other track estimators' digitized decisions are likely to be better than human judgment when the ball is not close to the critical zone; humans make big mistakes more frequently than machines make big mistakes.

But where the judgment is much more difficult—for both human and machine—then the potential error becomes crucial. To move forward it is necessary, first, to acknowledge that there is a distribution of error in the measurement and, second, to describe the characteristics of this distribution—wide or narrow. At worst, these should be described for the general case, but it would be much better to provide separate analysis for each of the main conditions that can affect the accuracy of the prediction. For example, it would be nice to know the distribution of error associated with fast balls and slow balls, with different lengths and ratios of the various trajectories, and, perhaps, with different condition of the pitch, the atmosphere, and the ball. For example, to what extent do track estimators make bigger errors in conditions when the ball tends to swerve as it travels through air after hitting the wicket?

The frame rate of the cameras will affect the accuracy of the prediction. Though the technical article referred to above suggests that Hawk-Eye takes its feed from standard broadcast cameras (which would have a frame rate of around 30 per second—"fps"), we will assume that the frame rate is 120 fps, as reported in 2004 on the Cricinfo website (see Questions 1 and 2 above).[5] In this case, if the ball is traveling at 80 mph (just under 120 feet per second), it would travel one foot between frames. This would make sense of Hawk-Eye's claim to need between one and two feet to make a prediction. In the worst case scenario, a ball traveling at 120 feet per second with a frame rate of 120 fps would need a minimum of two feet to provide three data points, which should be enough to estimate some kinds of constant curved trajectory, though there will obviously be some error and uncertainty. Fortunately, this lack of sure knowledge of frame rate and method of calculation does not affect the general principle of the argument, but it would still be nice to have an answer to Question 7:

Question 7: Does Hawk-Eye (or other track estimators) ever report a result based on only two data points? Do they directly measure postbounce swing? Do they measure the rate of change of post-bounce swing?

How Error Could Be Measured in the Case of Leg before Wicket

We do not know if data from the tests involving "thousands of deliveries" (see above) have been preserved. If they have, they might give an initial indication of the distribution of error as per Question 8.

Question 8: Have the data from the lbw test been preserved? Could these data be used to provide an initial indication of the distribution of error? Is it possible to make a more careful analysis of distribution of error?

But, of course, measuring the errors means being sure of where the ball actually went, and we don't know how this was measured and how accurate those measurements were:

Question 9: How was the point where the ball hit the stumps measured or recorded in the lbw test? How accurate are these measurements?

The beginnings of a more complete analysis could be made if, in a test like this, the cutoff point for the camera feed was systematically varied and the bounce point of the ball was

systematically varied so that a more complete range of potential errors was analyzed. In a still better test, these parameters,
and other parameters that affect the post-bounce behavior of
the ball, would be recorded and measured so that the likely size
of error could be reported for different circumstances. Some of
this may be beyond the capacity of companies who make track
estimators, but if it is then it should be clearly stated.

In an ideal world, our preference would be as follows. The
margin of error should be estimated from empirical tests, either
in the way suggested above or in some other way. Reports on the
method of testing and its degree of completeness and its possible unreliability should be made readily available. Subsequently,
using whatever knowledge of the distribution of measurement
error was available, confidence levels would either be associated
with Hawk-Eye and other track estimators' reconstructions in
real time or, if this were not possible, incorporated into the rules
that govern how track estimators are used. The graphic shown
to television viewers could be adjusted either to show something
like an error bar, or error circle, around the projected position of
the ball, or to indicate in some other way (such as numerical) the
confidence level for the prediction that wicket would have been
struck or not struck. Figure 3.5 (the putative errors are not drawn
to scale) indicates some of the possibilities, though no doubt it
could be improved upon.

If such changes were implemented, commentators might
remark, "The track estimator was 99.9 percent sure the ball was
going to hit the wicket so the umpire was right," or "The track
estimator was only 90 percent sure the ball was going to hit the
wicket—the umpire should not have given it out," or some such.
This way, not only would Hawk-Eye and other track estimators'
abilities be presented in a clearer and less easily misunderstood

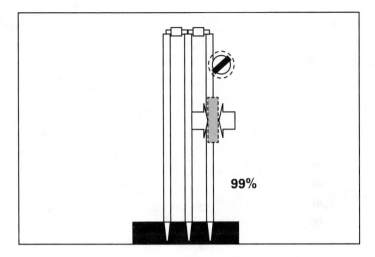

Figure 3.5
Some ways of indicating Hawk-Eye's possible measurement errors (not to scale).

way, but the technology itself could fulfill a valuable role in edu-cating the public about the way uncertainties are turned into decisions. What we are suggesting is not much more than is mentioned in the Hawk-Eye patent that we have found. There it is claimed that the HIT-MISS decision made by the apparatus is based on "whether the probability of the ball going on to hit the stumps is high e.g., above a given probability threshold" (8). We are asking for a more complete explanation of what the thresh-old is, what the probability is on any occasion, and for this infor-mation to be offered to the public. Nevertheless, we do think that the way track estimators are used in cricket is much more helpful than the way they are used in other sports (see chapter 5), particularly tennis, to which we'll turn in the next chapter.

4 Track Estimators and Tennis

Some people think the use of track estimators in tennis is radically different from their use in cricket because in cricket the track estimator has to project the path of the ball forward beyond the point at which it can make measurements, whereas in tennis it can follow the ball all the way to the line. Thus, in 2009, ex–England cricket captain and now cricket commentator, David Gower, wrote:

> [In cricket] we are not talking about the situation in tennis, for example, where Hawk-Eye is just tracking the ball and *showing fact in graphic form*. With an lbw referral, we have to acknowledge the element of computer-guided prediction as to where the ball would have ended up. That is an important difference.

But Gower is wrong. All cameras have a finite frame rate. If the frame rate is, say, 100 frames per second, and the tennis ball is moving at about 100 mph, it will travel about 1.5 feet between frames—more if the ball is traveling faster. The track estimator still has to estimate the track of the ball during those series of 1.5-foot-long elements between frames—after estimating the position of the ball from the pixels within single frames. Once more, if the ball's flight path is curving, this is still more difficult. Thus, the problem is the same in tennis as it is in cricket—though perhaps not quite so difficult, as in the case of lbw a longer path length unsupported by data points has to be estimated, which means that the potential for large error will be greater. The errors should be smaller in tennis than in lbw, but that does not mean there is no error.

In high-level tennis tournaments, players can challenge the umpire's decisions, and, if the player's challenge is supported by a track estimator such as Hawk-Eye, the original decision is overturned. In the summer of 2007 Hawk-Eye figured in two disputed line calls in which the ball was called out but, after a challenge from one of the players, was subsequently called in by Hawk-Eye.

Disputed Line Call A: Dubai

As reported on the Gulfnews website (May 3, 2009), in a match in Dubai between Rafael Nadal and Mikhail Youzhny, a challenge made by Youzhny was supported by Hawk-Eye:

World No 2 Rafael Nadal has questioned the efficiency of the new Hawk Eye line calling technology.

Thursday's first set between Nadal and Youzhny ended in a controversy with the tie-break score at 6–5 in favour of the Russian. Nadal thought a ball from the Russian had landed wide.

So the Hawk Eye was pressed into service and it showed the ball had skimmed the line.

But Nadal, chair umpire Roland Herfel of Germany and even Youzhny believed that the ball had landed wide after watching the Hawk Eye. But officials are bound to accept the Hawk Eye ruling.

"The mark of the ball was still on court and it was outside. But in the challenge it was in, so that's unbelievable. The Hawk Eye system is not perfect," fumed Nadal.

"I told the chair umpire: 'Look, the ball is out' and he said: 'I know.' ...

Even Youzhny agreed the ball appeared to have gone out. "I saw the mark, but I just challenged because it was a very important point," the Russian said.[1]

Disputed Line Call B: Wimbledon

Hawk-Eye was also central to a disputed line call in the Wimbledon men's final between Federer and Nadal. Nadal hit a ball that appeared to television viewers, to the umpire, and to Federer as impacting well behind the baseline, but Hawk-Eye called it in. Federer appealed to the umpire, but the umpire accepted the Hawk-Eye judgment. The following is the account from the *Telegraph* newspaper website, dated July 10, 2007:

Federer, a tennis conservative, has always been against the introduction of Hawk-Eye, and he was as angry as he had ever been on Centre Court when an "out" call on one of Nadal's shots was successfully challenged by the Spaniard in the fourth set. The Hawk-Eye replay suggested that the ball had hit the baseline; Federer thought otherwise. It was then that Federer asked umpire Carlos Ramos whether the machine could be turned off. Ramos declined but also seemed to suggest that he had thought the ball had landed long.

The Hawk-Eye review gave Nadal a break point, which he converted for a 3–0 lead, and Federer continued to complain during the change of ends. "How in the world was that ball in? S***. Look at the score now. It's killing me, Hawk-Eye is killing me," the Swiss said. So, a system which was introduced to prevent McEnroe-style rants at officialdom actually left one of the sport's gentlest champions fuming. (Hodgkinson 2007)[2]

Interestingly, this story also generated a lot of comments, many of which raise similar concerns to the ones made in this book, suggesting that at least some tennis fans do understand the concept of measurement error.

Technology and Tennis

In some ways the technical aspects of the tennis case are easier than the lbw case because the track estimator has more data points to go on, though we do not know if it uses postbounce data points in its calculations:

Question 10: In tennis, does Hawk-Eye use postbounce data points in calculating the bounce point?

Given that no combination of cameras provides an infinite frame rate, the computers still have to project forward (and back?) to generate the virtual trajectory. Again, we don't know what the actual frame rates are, so we don't know how much projection has to be done. Tennis balls can be served at up to 150 mph (ca. 220 feet per second), so in this respect the problem is worse for tennis than for cricket.

In tennis, there exists no traditional physical method for digitizing line-calls that is intrinsic to the game as with the bails in cricket. Decisions are traditionally made on the basis of fallible human observation—which is to say that the digitization is normally the preserve of the umpire's ontological authority supported by the epistemological privilege of the line judges.

In the case of tennis, we have some interesting analysis of the relationship between track estimators and humans, though it starts from the notion that track estimators set the standards of accuracy. George Mather, in an article written in 2008, has shown that humans can make large errors in judgment. From Mather (private communication) we learn that roughly 13 percent of 1,380 tennis challenges that he analyzed had a bounce point within 5 mm of the line according to Hawk-Eye and roughly 25 percent had a bounce point within 10mm. Let us reiterate that we are mostly concerned about the accuracy of track estimators in that 13 percent when the bounce point is close to the line; though it can be the case that large errors are sometimes made—as is probably true in the Nadal and Federer cases presented above—track estimators are probably better than humans most of the time when the bounce point is outside around 5 mm from the line. Whether they are *necessary* under these "big error" circumstances is another matter, since TV replays might do just as well.

Hawk-Eye reported in 2010 that the mean error in the position of the tennis ball as measured by its system is 3.6 mm and that the mean error for the pitching point in cricket was 2.6 mm. That the pitching point is the most accurate is no surprise since the trajectory from the arm of the bowler to the pitch of the ball is long and many data points can be gathered to aid in the reconstruction of the track. Let us also stress that we believe that such devices should be getting better all the time as technology improves and frame speeds increase, but again, we do not know the figures (see Questions 1 and 2 above). If we take the 3.6 mm (or whatever figure is current) as the mean deviation of the errors, we can do some back-of-the envelope calculations (but answers to Question 5 above would help).

This calculation rests on the assumption that the track estimator in question is not subject to systematic errors and that its random errors are distributed according to the well-known normal distribution we discussed above. This may not be the case for any particular track estimator but it the best we can do in the absence of publicly available measurements. If the mean deviation of an track estimator for, say, its measurement of the distance a tennis ball bounced in respect of a line was 3.6 mm, the standard deviation would be about 3.6 mm × 1.25 = 4.5

mm.[3] Because, in a normal distribution, 95 percent of the measurement lie within approximately 2 standard deviations of the mean and 99 percent lie within about 2.6 standard deviations, we can estimate some putative confidence intervals. In this case we could say that in 5 percent of Hawk-Eye's predictions (i.e., 1 in 20), the error could be greater than about 9 mm and in 1 percent it could be greater than 11.7 mm. The physics of the situation means that there could be an absolute upper cutoff point for the errors and this could be smaller than the calculation from an assumed normal distribution would imply, but we have no firm information as to whether this is the case. Even if the numbers we have calculated are correct, this would not mean that Hawk-Eye's call would be wrong every time it makes a significant mistake. This is because rightness and wrongness in terms of the binary decision (in or out) depends on the direction of the error. Nevertheless, if the figures are correct, it would be likely to be wrong on some of those occasions, and the Nadal and Federer incidents described above could have been such occasions. According to Hawk-Eye Innovations' own website, in the case of the Federer–Nadal call, Hawk-Eye called the ball in by only 1 mm: the possibility for mistakes is clear even if we look no further than the 3.6 mm mean deviation.

But what does the mean error of 3.6 mm in tennis imply? Is this the mean measured for all shots, including, say, lobs and low fast drives or serves? Just as in the case of varying kinds of ball in cricket, it seems likely that errors will not be equally distributed in tennis for different kinds of shots. For example, it is likely to be greatest in the direction of travel of a fast-moving ball. In this case, velocity across the line of travel is zero, or almost zero, but in the direction of travel, small errors in measurement will make a big difference to the position of the ball.

Question 11: Is the error for tennis shots greater in the line of travel of the ball than across the line? The mean error for tennis was at one time given as 3.6 mm—was that an average across all ball velocities and directions? If yes, does that mean that the mean error in the direction of travel is likely to be larger than 3.6 mm for the class of fast-moving balls?

In figure 4.1 the back line of a court is shown with a ball clipping the back edge. Uncertainty is indicated by the dotted circles. The diagram is only very roughly to scale at best, but the circle around the topmost ball is meant to show the error associated with between 2 and 3 standard deviations assuming a normal distribution and the other assumptions made above. In other words, on these assumptions, between 1 time in 20 and 1 time in 100 the actual position of the ball will be nestled up somewhere against the inside of the dotted circle.

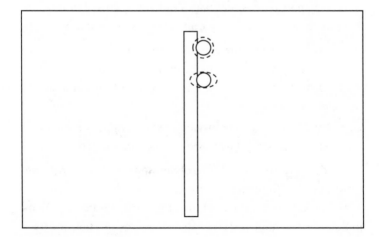

Figure 4.1
Is the error concentrated in the direction of travel of the ball and to what extent? (Not to scale.)

The lower ball shows roughly what the error would look like if it was concentrated into the direction of flight of the ball. Again, scale is not accurate and the degree of elongation of the oval might be exaggerated, but we cannot tell in the absence of more information. If the 3.6 mm mean is in fact averaged over all kinds of shots, it could be that in the case of fast drives or serves the elongation should be even more exaggerated. In fact, things are even more complicated than this: unlike cricket balls, tennis balls squash and distort when they hit the ground, so this has to be modeled by the track estimator before the "footprint" of the ball can be estimated. Likewise, the position of the lines is reconstructed within the track estimator, but this must also correspond to the relative position of the ball and line in the real world if the track estimator's graphic is to be an accurate representation of what actually happened. We are not in a position here to make a positive claim about these things; we are merely indicating possibilities that have not been discussed in the public domain as far as we can see and asking some questions:

Question 12: Has anyone compared the way the calculation of the way a tennis ball squashes and distorts with any measurements of what happens, and how close is the virtual image to measured results?

Question 13: What data points are used in Hawk-Eye's and other track estimators' virtual reconstruction of lines in tennis?

Incidentally, we were able to obtain some real error measurements, but not from Hawk-Eye.

Figure 4.2 shows an error distribution measured by Henk Jonkhoff, who built a system that measures the position of a tennis ball with wires buried in the court and a metallized ball; this system is known as Tennis Electronic Lines (TEL). Figure 4.2

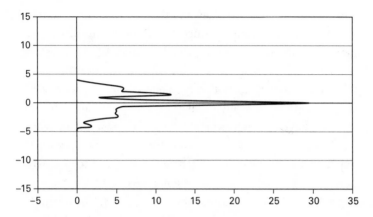

Figure 4.2
TEL's distribution of error.

is like figure 3.1—the graph showing the normal curve—except
it has been tipped on its side. If looked at as though one was
looking up while standing in the left-hand margin of the page,
it would follow the general shape of the normal curve—roughly
symmetrical with a high peak and roughly bell shaped (one
needs to use one's imagination somewhat). In this diagram the
"bell shape" shows that there were a lot of impacts of ball with
surface that deviated only a little from the true bounce point,
while as the deviation from the true bounce point—the measurement error—grows bigger and bigger in either direction, the
number of impacts affected by such an error decreases. In this
experiment, by the way, the true bounce point as defined by the
mark made in a film of talc on an alcohol-moistened Perspex
sheet—a very accurate way of gauging the error.

Of course, this is a real distribution, representing a relatively
small number of measurements, not a mathematical model of a
distribution representing an infinite number of measurements,

and that is why it is so spikey and ugly in comparison to figure 3.1. Also, we have only assumed that the distributions are normal in the absence of firm information, and they may well not be. But what this real curve shows is that, whether they follow the normal pattern or not, it is characteristic of error distributions that there are many small errors and fewer large ones, but there are large ones nevertheless. A number given for an average error, such as Hawk-Eye's 3.6 mm for tennis, 2.6 mm for the impact of a cricket ball on a pitch, and 5 mm for the lbw projection, tells you that there will be some larger errors, some much larger. The error distribution shown in figure 4.2 is the kind of thing we would like to see presented in the case of Hawk-Eye and the other track estimators, but at the very least, let us admit that there are errors of this kind and not pretend that track estimators always produce an exact result as is currently the impression given in tennis.

In an initial email exchange with Hawk-Eye Innovation's Tennis Operations Manager, we were referred to the International Tennis Federation if we wished to understand the methods of testing Hawk-Eye's errors. The International Tennis Federation (ITF) provides details of its testing procedures for automated line-callers on their website. We understand the ITF has the true position of the ball measured with very high-speed cameras. The crucial passages read as follows:

A3.4 Accuracy and Reliability

The decision-making success rate (i.e. "in" or "out" decisions) for all balls bouncing between 100 mm inside the line and 50 mm outside the line should be 100% with a tolerance of ± 5 mm.

The average absolute discrepancy for all impacts on a single line on court should not exceed 5 mm.

The maximum discrepancy between the system and the ITF for all impacts should not exceed 10 mm. ["ITF" seems to mean the ITF's measurement of the true bounce point. In an earlier draft this was described as a "true distance" but the ITF now seems to be accepting that all measurements, including their own, are subject to error.]

[In a footnote to the section, it reads:] The specifications described in this section apply to balls that legally cross a line from outside to inside.[4]

We found these rules difficult to understand. Initially, we could not understand how in/out decision making can be 100 percent accurate if there is a tolerance of 5mm. On the face of it, these statements seem incompatible—a ball could be 5mm out and still be called in. Even if we forget about distribution of error and just accept the 5 mm at face value, if Hawk-Eye was taking its measure of accuracy from the ITF, the Federer–Nadal disputed ball might well have been out by nearly a quarter of an inch, even though Hawk-Eye called it in.

Question 14: In the case of tennis, does Hawk-Eye (or any other track estimator) utilize measures of accuracy in addition to those demanded by the ITF test? If yes, what are they?

If we accept Hawk-Eye Innovations' own figure of 3.6 mm average error and its claim that the ball was 1mm in, the possibility for a mistake is still obvious. The ITF appeared to agree; in response to our inquiries (all of which took place on January 26, 2008), their spokesperson said:

... In general, if the ball landed sufficiently close to the edge of the line, there is a chance that Hawk-Eye could make the wrong call. (private communication)

Hawk-Eye Innovations' own website contains a discussion of this specific line call. The introductory paragraph remarks:

This document provides more information about the line call that Roger Federer questioned during the Wimbledon Men's Singles Final on Sunday 8th July. Whilst it is unable to prove conclusively that the ball was 1mm "in" as shown by Hawk-Eye, it can show that 1mm "in" is a likely [*sic*].[5]

The ITF was also able to clear up at least some of our confusion in a speedy way. Here is the gist of the initial response from the ITF (private communication):

All decisions made by a line-calling system ("in" or "out") must be correct, unless the ball lands within 5 mm of the outside edge of the line, when an incorrect decision is allowed, providing that the absolute error in the system's measured impact location is no more than 10 mm.

Example 1.
True impact location: 4 mm "out."
System's measured impact location: 2 mm "in."
Outcome: Acceptable (wrong decision, but absolute discrepancy < 10 mm).

Example 2.
True impact location: 4 mm "out."
System's measured impact location: 8 mm "in."
Outcome: Unacceptable (wrong decision, absolute discrepancy > 10 mm).

In sum, the ITF accepts errors of up to 10 mm for individual impacts, and the system may still pass the accuracy test overall (contingent on meeting the other performance criteria).

Incidentally, we asked the ITF how many impacts were involved in their tests. They explained:

Over the full evaluation, at least 80, and normally 100–120. Of these, around 10% land within 5 mm of the line. (private communication)

Thus, it could be that the ITF tested Hawk-Eye's performance in the crucial zone around the edge of a line on only around 6 to 15 impacts of ball with court.

To conclude on random error in tennis, it seems to us that the contribution of Hawk-Eye would be much better understood if, just as in cricket, it were admitted that on a few occasions it will

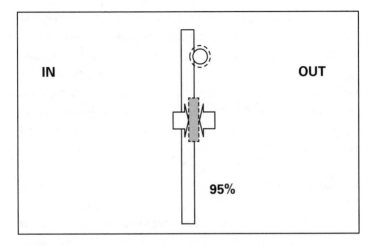

Figure 4.3
Some ways of indicating possible error in line calls in tennis (not to scale).

be wrong and if each prediction were associated with a confidence interval. Each line call provided by Hawk-Eye in real time should be associated with a claim about the confidence level of that call. Figure 4.3 suggests ways in which these possibilities could be indicated to the public (though the error circle might need to be elongated as in figure 4.1).

As time has gone on we realize we are probably asking too much of the people involved in sports, who are not primarily scientists and are driven by values more to do with profit and entertainment than a burning desire to deliver the truth of the matter with the maximum attainable clarity. Nevertheless, even if these groups do not have the skills, resources, or desire to make the kinds of measurements and the kinds of presentations to the public we would like to see in an ideal world, they do have the skills and resources to be able to say when a device is not as accurate as it appears to be. We have quoted the International Tennis Federation as saying that it is acceptable to them if a track estimator calls a ball 4 mm out when it is 2 mm in, and we have quoted Hawk-Eye's makers as saying that their devices have average errors of 3.6 mm in tennis (maybe a little better these days); and yet they seem reluctant to make clear to the sports-viewing public the fact that these devices make errors—quite the contrary. Why?

The Skidding Ball

We have looked at random error, but systematic error is more complex. The Hawk-Eye Innovations website makes the claim, supported by stills from a high-speed camera, that the human eye, and television replays, can be systematically misleading under certain circumstances.[6] The ball may just touch the line but skid so that it is still in contact with the ground when it bounces upward well beyond the line, giving the impression that it did not, in fact, touch the line at all—the eye will see a space between ball and line when the ball bounces upward. The website seems to show, then, that it is possible for the ball to appear out on a television replay or to the naked eye but still be just in, in the sense that the ball did just touch the line before its apparently out position became revealed to the eye or the camera. Barring unknown sources of human or machine malfunction, the disagreement between humans and Hawk-Eye in the case of the Wimbledon dispute has, then, two possible explanations. The first is the one given in the preceding sections, namely that it was a random measurement error in Hawk-Eye. The second, made at least feasible by Hawk-Eye Innovations' analysis of the skidding ball, is that the disagreement results from a systematic error in human judgment.

But this business of the skid and bounce making a ball appear out when it is really in is a symptom of the misplaced fetish with accuracy that is one of the themes of this book. During the whole history of tennis before track estimators, and in the case of every tennis match played today without the benefit of track estimators, a ball that just touched the line but skidded and bounced such as to appear out is a ball that *is* out, not in. It is the umpire's ontological authority that has always made it so, and there no injustice has been done, because to every human and television viewer it appears out. Prior to track estimators, for TV viewers with the benefit of replays, the call of "out" for such cases was a case of *transparent justice*. Track estimators and Hawk-Eye's website now want to claim that this huge number of calls were really cases of *false justice* because in all these cases the ball was really in, even though no one could see it and everyone was perfectly happy. And furthermore, for all those games played in the park and at the club where track estimators are not used, the players and viewers are condemned to false justice for the rest of time.

This is madness. It is double madness because in the case of a tennis ball and a line painted on grass we do not, and cannot, know exactly what "in" and "out" mean. Why not just leave this decision as it has been throughout the ages and is in every non-track-estimator-assisted game? If the ball appears out to every human, however sharp their eyesight and irrespective of how many TV replays they observe, then it is out. That is what follows from being concerned with justice and the viewer, rather than enthralled by technology and a fetish for a mythical account of accuracy; and madness is what follows from the accuracy fetish.

5 The Impact of Track Estimators

It might be that Hawk-Eye or the equivalent could become *defined* as the "decision maker" rather than a "decision aid" so that questions concerning its accuracy would no longer be relevant. Instead of estimating where the ball landed, the track estimator would be the creating the fact about where the ball landed. In other words, ontological authority would be transferred from the umpire to the track estimator; the track estimator's reconstruction would fulfill the same definitive role as the bails in cricket. This is already beginning to happen in tennis, though at the moment it occurs only when a player formally challenges a call.

Making the track estimator the authority in this way could resolve a number of problems by providing readily acceptable explanations for otherwise borderline decisions. No one claims the bails are inaccurate because everyone accepts that a dislodged bail is the digitized definition of "hitting the wicket." It would be the same if the track estimator's decisions were to be made the defining criterion of lbw—setting aside the need for the umpire to make the prior decision about whether the batter was playing a stroke. A track estimator could also be taken to *define* "in" and "out" in tennis. Players would come to talk in terms of "bad luck" if a call went against their own judgment rather than "inaccuracy," just as they now talk of bad luck rather than inaccuracy if a bail does not fall when the ball touches a stump too gently. Indeed, in tennis, players nowadays seem to have accepted the track estimator with resignation and now try to "game it" by always using up their challenges toward the end of a set just in case they get lucky. How would using a track estimator as the definer of lbw and "in" and "out" change things— and are these changes already happening?

The first change would be a good one because it would mean abandoning the accuracy fetish; one would no longer care if the track estimator was right or not, one would only be concerned with what it decided. Of course, this would pull top-level sports away from the game the rest of us play, because the rest of us don't have track estimators and would have to make our own decisions. Now let us look at other possibilities.

In cricket it could mean substantial changes. For instance, it has been argued that if Hawk-Eye's "face value" lbw projections were taken as the defining criterion, many cricket games in which it was used would be much shorter, perhaps leading to a financial crisis. As discussed above, in lbw decisions made by umpires, a systematic error is deliberately introduced via the well-established rule that the batter gets the benefit of the doubt. Since there is a lot of doubt in human lbw decisions, it is often quite rare for a decision to be made in the bowler's favor. It has been suggested that Hawk-Eye's projections, if taken literally, would greatly increase the number of lbw decisions unless batters started to play differently.

Question 15: Has any systematic analysis of human and Hawk-Eye lbw calls been made? If yes, does it show that the number of lbw decisions given "out" would go up if Hawk-Eye's decisions were applied automatically? If yes, by how much?

Hawkins is quoted by S. Rajesh (2004) as saying that one in five lbw "out" decisions are incorrect and Hawk-Eye could remedy this. But it seems unlikely that this would balance out the increased number of outs consequent on loss of benefit of the doubt, though a more complete analysis is desirable. What we do know is that cricket commentators nowadays are certain that many more lbw decisions are given "out" when the batter is well forward and that this has changed the way spin-bowlers bowl, batters move, and umpires make their decisions even when track estimators are not in use; cricket umpires seem far more ready to give a batter out when he or she is well forward.

Something not dissimilar appears to have happened in baseball since cameras have been introduced that can track a pitch and record it automatically as a strike or a ball. These are not used in the games themselves, unlike TV replays, which are used for every decision except the ball/strike decision, but they are used to judge umpires' performance; umpires can now lose their jobs if they make too many mistakes and can earn financial rewards if they do especially well in terms of the accuracy of their calls. The result is that the bottom edge of the strike zone has moved downward and batters are more often beaten by low strikes that would once have been called balls; this has reduced the number of home runs and, some would argue, has also reduced the attractiveness of the game. An article describing this can be found on the Web, indicatively entitled "The Simple Technology That Accidentally Ruined Baseball."[1]

As we've intimated, similar considerations apply in tennis. Given that we are now conscious of the skid and bounce, umpires and players should be less confident in stopping a rally when the ball appears out to the naked eye.

Question 16: Has any systematic analysis been made of changes in patterns of play or decision making since track estimators came into use? If yes, does it show any systematic bias, and, if so, what is the direction and degree of the bias?

It should be noted in passing that the question of whether track estimators would bring about such a change is now being obscured. Because we have become convinced that TV replays do not tell the truth in the case of the skid and bounce—or for other reasons to which we are not privy—we no longer see TV replays of close calls in tennis. Thus, the viewer has nothing to go on regarding whether the track estimator was right or not, and that includes cases where the mistake might have been glaring enough to be obvious on TV and not subject to the subtle objection in respect of a skidding ball—for example, it includes cases where the mistake is in the sideways direction, not the direction of ball travel. This seems an unnecessary restriction, and allowing viewers to see both the normal replay and the track estimator's reconstruction would allow for a more informed debate about which kind of decision maker is preferable. As it is, this debate has been closed off because only one set of evidence is available.

How Should We Use Sports Decision Aids?

Our original paper containing much of the above analysis was published in 2007. In it we also made a recommendation for how to retain the "benefit of the doubt" rule in the case of lbw alongside the use of track estimators. We argued that the

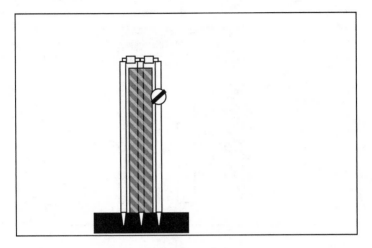

Figure 5.1
Possible ways to reproduce human systematic error (not to scale).

effective size of the wicket should be reduced as far as the track estimator was concerned, and we illustrated the idea in a figure, reproduced here as figure 5.1.

What we said was that "to reproduce the 'benefit of the doubt' rule it would only be necessary for Hawk-Eye to make its decisions on a smaller virtual set of stumps indicated by the shaded box. Balls predicted to hit the wicket in the area outside that box would count as 'NOT OUT' on the basis of benefit of the doubt." Subsequently, whether by coincidence, as a result of advice from others, or as a result of our efforts, the International Cricket Council, who had certainly requested a copy of our paper, adopted a similar idea. They concluded that a track estimator decision should not overrule an umpire's decision unless the track estimator's estimate showed that at least half the ball was inside a line drawn at the center point of the outer stump.

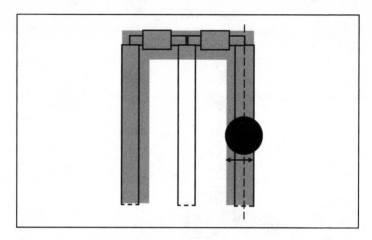

Figure 5.2
The gray box illustrates the zone of uncertain on one edge of the wicket
equal to half a stump and half a ball (not to scale).

Figure 5.2 shows the estimated striking point of a ball needed to
overrule an umpire's decision.

This creates what we are going to call "the zone of uncertainty," which applies to the top of the stumps—was the ball
hitting or going over?—as well as the sides. A cricket ball can be
up to 22.9 cm in circumference = roughly 7.3 cm in diameter
while a stump may be up to 3.81 cm in width, so the zone of
uncertainty is 55 mm (half the width of the ball plus half the
width of one stump) around the edge of the stumps.

What happens in the case of lbw in cricket is that the
bowler and other fielders appeal to the umpire: they shout out
"Howzat?"—a shortened form of "How was that, umpire? We
believe it was leg before wicket," which is usually executed as
a tortured scream with lots of jumping up and down and arm
waving, and even repetition (which is frowned upon), intended

to impress the umpire with the sincerity of the bowler's and fielders' beliefs. The batter, meanwhile, unless the matter is absolutely clear, will adopt a nonchalant attitude and a quizzical expression—"What are these fools shouting about?" The umpire will then give a decision: a raised finger indicates "out," an almost imperceptible shake of the head or no movement at all indicating "not out."[2] At this point in the modern game, the players have a right to challenge; there may be a short consultation between the two batters, or between the captain, bowler, and wicketkeeper if the decision has gone against the fielding side, in which they decide whether to use up one of their limited number of opportunities to challenge the call. They have three challenges per test-match innings, which are used up one by one when they are unsuccessful though unaffected if successful. If they decide to challenge, the sequence of events that unfolds in cricket is interesting and instructive.

First officials will use a TV replay to ensure that the bowler did nothing illegal, like stepping over the crease when the ball was delivered in which case it would be counted as a "no-ball" and the batter cannot be given out. Next they will use other devices, including Hot-Spot and Snicko (and possibly TV replay again), to make sure the ball did not touch the bat before hitting the pad. If it didn't, officials will then turn to the track estimator.

The track estimator's graphic is used to make three decisions in sequence. The first decision is whether the ball pitched on the rectangular strip between wicket and wicket or if it pitched to the leg side of the strip. Remember, the batter cannot be out by lbw if the ball pitched outside the strip on the leg side (the umpire will already have made the decision about whether the batter was playing a stroke). The second decision is whether the batter's pad, at the moment it was struck with the ball, was between wicket and wicket. In these two cases, the zone of uncertainty is simply half the width of the ball. The third decision is whether the ball would have gone on to hit the stumps, where the zone of uncertainty is half a ball plus half a stump. Each of these decisions is presented in sequence with traffic-light colors indicating their significance. A single green means the batter is safe; three reds mean the batter is out. But a single yellow means the ball was striking either pitch, or pad, or stumps, in the zone of uncertainty, *and the original umpire's decision stands!* Let us call this the "Right If Not Wrong" (RINOWN) principle. The Right If Not Wrong principle means the umpire is right unless the track estimator shows, unambiguously, that he or she was wrong.

It is that italicized phrase that is the key to whole system we are recommending in this book. The way decision aids should be used in all sports is as they are used in cricket's lbw decisions: "the match official's decision stands unless that decision can be shown quickly and easily to be clearly wrong." That means that in all sports where such decision aids are used—and decision aids include TV replays—match officials ought to make a decision every time, and not call for the use of any technological aid until they have made it. They ought to take responsibility as they do in every match not supported by complex technology; they should retain and celebrate their ontological authority. The match official's decision is then "the decision," unless it is agreed that, using a technological aid, it can be clearly shown to be wrong. Thus, the RINOWN principle resolves the justice problem: if the technological aid quickly and clearly shows the match official is wrong, then the TV viewer will know it too. If the technological aid cannot show quickly and clearly that the match official is wrong, then the TV viewer, just as in the long tradition of sports, has nothing to complain about and simply has to accept that the match official's epistemological privilege puts them in the best position to make a good decision. They must concede the ontological authority to the official.

The Right If Not Wrong principle is similar to the practice in American football. In American football, coaches can make a limited number of challenges to match officials' calls and TV assistants can also choose to review a call. The rule is: "The referee has 60 seconds to watch the instant replay of the play and decide if the original call was correct. The referee must see 'incontrovertible visual evidence' for a call to be overturned." In other words, the on-field officials are right unless there is incontrovertible evidence that they are wrong. The technology used to provide the evidence is TV replay.[3]

In our language, the zones of uncertainty and the way they are used accomplish a number of tasks. First, they take explicit account of the uncertainty associated with the track estimator technology and have the potential to convey that uncertainty to viewers. Unfortunately, because of the way track estimators present their reconstructions, their implications are not always clear. Thus, the ex–England cricket captain, Ian Botham, now a regular broadcast commentator, is often heard to complain that cricket's use of track estimators is unfair and inconsistent because the same reconstructed ball-flight will lead to either "out" or "not out" depending on the umpire's initial decision. Botham thinks the same reconstruction should always lead to the same outcome—this is what he thinks of as consistency. But Botham, probably misled by the very exact graphic presented by Hawk-Eye and other track estimators, with their sharp lines and perfectly defined ball, misses the fact that though the graphic is exact the real ball might be in a number of different places—that's what error means. So when we see a similar graphic representing two different incidents, it does not mean that what happened in the two different incidents was the same, and that is why there is no inconsistency. To resolve this tricky problem, the graphic of the ball should be large and fuzzy, representing a range of positions where the ball *might be* according to the track estimator reconstruction. A fuzzy graphic would remove the sense of inconsistency.

Second, the zone of uncertainty, together with its 55 millimeter tolerance, defines what is to count as a big error. If the ball is shown to be hitting the wicket inside this zone, there is no big error, and the on-field umpire retains epistemological privilege by virtue of experience, viewpoint, and training. In contrast, where the track estimator shows the ball passing clearly within or completely outside the zone of uncertainty, the on-field umpire is overruled and the original decision overturned. Under these circumstances we agree that the umpire might have made a big error spotted by the track estimator. Once more, however, if the errors are big enough, there is also less reason to think that TV replays could not correct them without needing track estimators.

Third, if the technology is used properly, delay is minimized: a decision is made quickly, just as it would have been in the traditional game and games that do not have the benefit of technical assistance—that is, most games—and the technological device can be consulted right away because one is looking for clear mistakes only, which can be spotted quickly or not at all. A change for cricket that could follow from these arguments is in the way umpires handle stumpings and run outs. As it is, when players appeal for a stumping or run out the umpire tends immediately to refer the question to the "third umpire," who will then inspect and broadcast TV replays to decide the matter. Sometimes TV replays are not unambiguous where such decisions are involved, and delays and uncertainties can result. If cricket wants to speed up these decisions, or if cricket wants to accept our philosophy and return as much ontological authority to the umpire as possible, then umpires should always make their best decision—"in" or "out"—before referring the matter. Then, when there is any ambiguity, the decision-making process can be cut short—the umpire's decision was right unless it can be quickly and easily shown to be wrong: RINOWN! The same set of considerations applies to other sports, such as rugby union. Here, in the case of a disputed "try," referees tend to go straight to third officials for a decision. We think they should make their decision first—just as in the traditional game and every game played without complex technological aids—and only then call for a TV replay. Once more, RINOWN should apply and the original decision should stand unless the TV replay shows quickly and clearly that it was wrong. This would speed the game and bring it much more in line with all the other games of rugby being played in schools and clubs.

Track Estimators in Tennis Once More

Going back to tennis, as we have seen, in the case of a challenge the ontological authority is delegated to the track estimator, and the idealized track estimator graphic reply indicates that it is capable of making single-millimeter decisions regarding whether a ball was in or out. Our argument is that this is misleading and that track estimators cannot do what they appear to do; furthermore, that the problem is justice, not accuracy, so that millimeter accuracy is not needed. Makers of other sports decision aids, such as TEL and other track estimators, have told us that it is a mistake to promote such devices on the basis of accuracy alone, as it is impossible to eliminate all errors. Instead, these manufacturers see the technology as solving a different problem, namely avoiding prolonged disputes between players and umpires. We, in turn, see it as a question of bringing epistemological privilege back into harmony with ontological authority. In either case, if the goal is to avoid big mistakes, then simply showing that the ball bounced very close to the line is enough to show that, as in cricket, the umpire's decision was reasonable.

The obvious solution is to adopt the approach that has proved successful in cricket—define zones of uncertainty and apply the Right If Not Wrong principle. It would seem to us that, given the average errors reported and bearing in mind the way errors are generally distributed, and taking into account the International Tennis Federation's definitions of acceptable performance by a track estimator, which allows errors of up to 10 mm, the zone of uncertainty in tennis should be at least 5 mm. If a track estimator reconstructed the distance between ball and line as 5 mm or less one way or the other, then the umpire's decision should stand. To take account of less than optimum conditions, the "tails" of the statistical distribution, and the blurred edges of the lines themselves it might be wise to make the zone wider, though by how much is a matter for research and experience. Figure 5.3 illustrates the zone of uncertainty in tennis though not the exact width of the zone.

As with cricket, this would return ontological authority to the umpire except where a large mistake—that is, where the track estimator shows the trailing edge of the ball to be outside

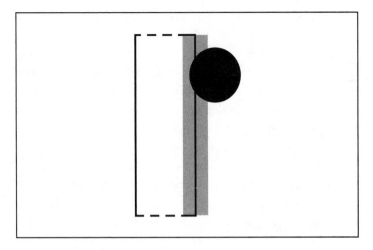

Figure 5.3
Zone of uncertainty on the crucial edge of a tennis line: if the trailing edge of the ball is shown to be in the zone, the umpire's initial decision stands (not to scale).

the zone of uncertainty—is seen to be made. This is RINOWN. And, once more, it would also go some way to realigning the technologically assisted game with the rest of tennis as, presumably, redefining skidding and bouncing balls as "in" because they touched the line in a way that was inaccessible to human perception would no longer happen: justice, not accuracy, would rule.

We believe, however, that more research is needed to compare how this would work with TV replays and track estimators. We think TV replays should be brought back into the game so that the viewer could see how judgments based on TV replays compared with track estimator judgments. It may be that with a zone of uncertainty, they would reach the same verdict.

These changes would remove what can only be called the current scandal, where, in the case of close calls, the crowd is bemused by what is essentially a video game while believing they are seeing what really happened. Presentation of track estimator results alongside TV replays would, on the other hand, reassure viewers that justice was being done in these cases and put an end to the embarrassment of having both players and umpire agree that the ball was miscalled by the track estimator but still having to accept its decision. Should it turn out that TV replays and track estimators were always in agreement when the bounce was outside the zone of uncertainty, it may be that track estimators could be dispensed with altogether in respect of decision making.

The Value of Track Estimators

Strange though it may seem, in spite of the difficulties that the authors of this book have encountered with some of the sports-decision-aid companies, and in spite of our skepticism in regard to their accuracy, we nevertheless believe that track estimators have hugely enriched sports and we hope they will continue to do so. The excellent graphics produced that, in cricket, show the distribution of scoring shots for particular players (the "wagon wheel") or the tactics used by different bowlers (the "beehive" and the pitch points) enhance the viewing experience and contribute to the spectacle and enjoyment of the game. The same could be done for tennis (or has already been done; see appendix 4). Using these graphics, tennis fans would be able to understand the different way different players play the game—which parts of the court they exploit and which they do not, for example. We visualize that, over time, track estimators could build up a rich archive of these sports and others. An Internet search could be able to reproduce the career pattern of shots of a tennis player or a cricket batter or enable the searcher to compare the different pitch points or degrees of spin imparted over a career by any pair of bowlers. The key point is that this kind of use of track estimators does not depend on millimeter accuracy and therefore will not mislead.

We now turn to the Premier League, and look at how it might have turned out with the use of TV replays.

6 How the Premier League Might Have Turned Out with TV Replays

We now return to the question of justice and heartbreak with which we began the book. Is it the case that in the English Premier League referees' mistakes "all even out in the end," and, if they do not, how could the situation be put right? We answer that things could be put right with the use of TV replays.

We now show what the Premier League might have looked like had TV replays been used to aid referees over three seasons: 2011–2012, 2012–2013, and 2013–2014. We show that the winners would probably have been different in two out of three seasons, that in one season a different team would probably have played in the prestigious European Champions League, and that in one season a relegated team would probably have gained two places and stayed up while another team would have dropped five places and gone down.

What is the status of these claims, and what does that word "probably" mean? It means that if we had to bet on the outcome of those seasons given perfect refereeing, that is the way we would have bet. We might not have won every bet, but it is still certain that something would have been significantly different even if these exact teams were not affected in this exact way. We can demonstrate this by thinking about the numbers that give rise to positions in the league tables. Consider that in every football season there are twenty positions in the Premiership at stake, which means there are nineteen sets of points-differences between teams: the difference between first and second, the difference between second and third, and so on, all the way down the difference between nineteenth and twentieth. In English football, zero points are awarded for a loss, one point for a draw, and three points for a win. Now, each season involves 380 games: twenty teams each play nineteen other teams home and away. In the 2012–2013 season, in 262 of these 380 games (69 percent) the difference in goals scored by the two teams was 1 or 0; in 2013–2014 the figure was 216 (57 percent), and in 2014–2015 it was 247 (65 percent). That means that in these three seasons—and the numbers are pretty constant over the years—in half to two-thirds of the games played, a single goal disallowed, or a single goal awarded instead of disallowed, would have turned a draw into a win or a win into draw for one team and a loss into a draw or a draw into a loss for the other team; one team would have gained or lost two points and one team would have gained or lost one point.

What is the impact of gaining or losing one or two points? We can begin to answer this question by looking at the points-differences between teams in different positions in the final league table. Table 6.1 shows the results for the eleven years up to and including 2014–2015. With twenty teams there are nineteen points-differences between the team above and the team below. Row 1 of the 2014–2015 column shows that in that season, in two of those pairs of positions, the points of the upper and lower team were the same and the teams were separated only by "goal difference" (goals scored minus goals conceded). Row 2 of that column shows that two of the positions were separated by only one point. Row 3 of that column shows that five of the positions were separated by two points, and so on all the way down the column. The other ten columns show the outcomes for ten other years. The total of nineteen differences in every year is confirmed in row 20 of the table. Row 21 of the table shows how many positions in each year were separated by two points or fewer. As can be seen, it is more than half of the positions in every year except 2014–2015. That means that, in nearly every year, more than a half of the positions in the table would have been swapped with the one below or above as a result of a gain or loss of one or two points. And we know that such a difference could have arisen out of the scoring or not scoring of one goal in around two-thirds of the games played. One goal can make a lot of difference.

Now let us turn to the "crucial" positions. There are three crucial points-differences per season: the difference between first and second, the difference between fourth and fifth (with fourth going to play in the prestigious and financially rewarding Champion's League and fifth playing in the European Cup, which involves lots of games for little reward), and the difference between seventeenth and eighteenth, since the eighteenth, nineteenth, and twentieth teams are relegated from the Premiership to the much less financially rewarding lower leagues. The results for the crucial positions are set out in rows 22–24 of the table. As indicated by the shaded cells, in every year except 2014–2015 at least one of those positions would have been affected by a change of two points or fewer in the aggregate points-score of one team and in three of the years two of the positions would have been affected. In five of the years a different team would have won the Premiership, and in six of the years a different team or teams would have been relegated as a result of a change of two points or fewer. So one or two points will often affect a crucial position, and that means a single goal will often affect a crucial position.

To repeat, what table 6.1 tells us is that there is every chance that a point or two could affect a crucial position in the Premiership, and we know from the earlier discussion that this kind of difference can be brought about by the awarding or not awarding or a single goal in a half to two-thirds of Premiership games. It seems, therefore, extremely unlikely that it will all "even out in the end." Given these numbers, that it will all even out in the end would be a miracle.

But we wanted to get a better idea of what difference could actually have been made, even though there is no method that will guarantee absolute accuracy. We decided to reexamine most of the matches played between 2011 and 2014 and rescore them after correcting referees' mistakes. There are 380 games per season, and we looked at 347, 348, and 354 games in the three seasons we analyzed. The games we analyzed were all those we could find discussed on the BBC program *Match of the Day* (*MoTD*). This program employs a panel of pundits—ex-footballers—to go through edited versions of the matches on the evening after they have been played, and, unsurprisingly, the main talking points are goals and contentious referee decisions. The contentious decisions are examined in detail using multiple replays; where "offside" is in contention, computer graphics are used to transform the frame showing the moment the ball was passed forward so that viewers can look along the line of the most forward players. We don't know how exactly how accurate this method is, and, as explained earlier, we could not find out because the BBC did not respond to our requests for information, but we have no a priori reason to question it.

We now had as a resource the ex-footballers' comments and conclusions, and, by assessing the meaning and certainty of their remarks—a standard procedure in social science research, which often uses interviews to gather data—along with our own view of the incidents aided by multiple replays and electronic enhancement, we formed a view on what the decision should have been and how certain we could be about this judgment. In cases where we thought the referee had made a mistake, we then had to make an estimate of its effect on the game—what would the outcome have been if the referee had made a different decision.

Trying to work out how an outcome would have changed if things had been different is known as the method of the "counterfactual," and it is fraught with danger. Maybe the change would have affected not just the score but also the attitude of the players—they might have played the rest of the game in a different manner, with more motivation or with their spirit broken; the referee's subsequent decisions might also have been affected. All we can say is that our conclusions about the effect on the game would likely be fairly similar to the conclusions of the teams involved in the match. One is not going to find teams saying, "I know we lost by one goal, and it is true that Malleus should not have been sent off because his arm was nowhere near Incus's face in spite of all that rolling around," or "I know we lost by one goal and that penalty awarded against us was a disgrace because Stapes dived," and then going on to say, "but we may have lost anyway—who can tell when counterfactuals are involved?"

The change in score and the change in points that we assigned were affected by a weighting that covered both our certainty that the referee had made a mistake, the state of the game, and

Table 6.1
Number of various points-differences in the Premiership table over eleven years, from 2004 to 2015

#	Points Diff	2014-15	2013-14	2012-13	2011-12	2010-11	2009-10	2008-09	2007-08	2006-07	2005-06	2004-05
1	0	2	1	3	5	5	2	5	1	2	5	4
2	1	2	4	4	4	5	5	4	5	5	5	5
3	2	5	5	4	2	1	3	2	6	5	6	1
4	3	4	4	4		2	3	1	2	1		4
5	4	2	1		2	2		2		2	4	
6	5	1	1		3	1	3	2	1		1	2
7	6	2	1		1	2			1	2		1
8	7		1						1			
9	8	1	1	1	1					1	1	
10	9			1		1		2				
11	10						1					
12	11			1			2	1	1			
13	12			1								1

	Diff	2014-15	2013-14	2012-13	2011-12	2010-11	2009-10	2008-09	2007-08	2006-07	2005-06	2004-05
	YEAR Points											
14	13											
15	14											
16	15									1	2	
17	16											1
18	19				1							
19	24								1			
20	Total	19	19	19	19	19	19	19	19	19	19	19
21	0/1/2	9	10	11	11	11	10	11	12	12	11	10

	Position	2014-15	2013-14	2012-13	2011-12	2010-11	2009-10	2008-09	2007-08	2006-07	2005-06	2004-05
	Points-Differences between Crucial Positions											
22	1:2	8	2	1	0	9	1	4	2	6	8	12
23	4:5	6	7	1	4	6	3	9	11	8	2	3
24	17:18	3	3	3	1	1	5	1	0	0	4	1

the point in the game at which the mistake was made. Obviously a mistake that wrongly awarded a penalty to a team that won or lost by two goals or more would not be counted as making a difference; a penalty awarded at the beginning of a close game would count less than one awarded at the end of close game; a mistaken sending off at the very end of a game would count less than a mistaken sending off at the very beginning of a game; and so on. If two incidents occurred, one favoring one team and one the other, they would generally be taken to cancel each other out. There is no formula for these judgments—they are just judgments made by someone who knows and loves football and has watched through three seasons of *MoTD*.[1] To repeat, there are plenty of ways to go wrong because lots of judgments are involved, and we have looked at only about 92 percent of the games instead of 100 percent. This means that if we could travel to an alternative universe in which the same games were played in exactly the same conditions but with referees who were all-seeing and all-knowing, the outcomes of the seasons might not be exactly what we intimate below. But that doesn't matter too much for the argument of the book: the argument of the book is only that something importantly different would have happened even if the referees had not made exactly these obvious mistakes, If what happened wasn't exactly what we concluded, that's OK. Of course, we have tried our best so we still think we are right, or very close to right, about these three seasons.

Tables 6.1–6.3 show the league tables as they were and what they would have looked like had they been adjusted according to our *MoTD*-based method of correcting refereeing mistakes. The original table is shown on the left with the adjusted table on the right. Crucial positions that change are shown with

Table 6.2

2011–2012

Pos	Actual table	Pts	Adj	Adjusted table	Pts
1	Man City	89	0	Man Utd	95
2	Man Utd	89	6	*Man City*	89
3	Arsenal	70	0	Tottenham	71
4	Tottenham	69	2	Arsenal	70
5	Newcastle	65	−1	Newcastle	64
6	Chelsea	64	−4	Chelsea	60
7	Everton	56	−1	Liverpool	58
8	Liverpool	52	6	Everton	55
9	Fulham	52	0	Fulham	52
10	WBA	47	−4	Swansea	46
11	Swansea	47	−1	Norwich	46
12	Norwich	47	−1	Wigan	46
13	Sunderland	45	−4	WBA	43
14	Stoke City	45	−10	Aston Villa	42
15	Wigan	43	3	Sunderland	41
16	Aston Villa	38	4	QPR	37
17	QPR	37	0	Blackburn	35
18	Bolton	36	−2	*Stoke City*	35
19	Blackburn	31	4	Bolton	34
20	Wolves	25	9	Wolves	34
	Totals	1,047	6		1,053
	Points Moved		62		

Table 6.3

2012–2013

Pos	Actual table	Pts	Adj	Adjusted table	Pts
1	Man Utd	89	*1*	**Man Utd**	90
2	Man City	78	*2*	**Man City**	80
3	Chelsea	75	*3*	**Chelsea**	78
4	Arsenal	73	*–6*	Tottenham	67
5	Tottenham	72	*–5*	*Arsenal*	67
6	Everton	63	*3*	**Everton**	66
7	Liverpool	61	*3*	**Liverpool**	64
8	WBA	49	*–4*	**Norwich**	51
9	Swansea	46	*0*	**West Ham Utd**	47
10	West Ham Utd	46	*1*	**Fulham**	47
11	Norwich	44	*7*	**Swansea**	46
12	Fulham	43	*4*	**WBA**	45
13	Stoke City	42	*1*	**Stoke City**	43
14	Southampton	41	*0*	**Southampton**	41
15	Aston Villa	41	*–3*	**Newcastle Utd**	41
16	Newcastle Utd	41	*0*	**Aston Villa**	38
17	Sunderland	39	*–5*	**Sunderland**	34
18	Wigan	36	*–3*	**Wigan**	33
19	Reading	28	*4*	**Reading**	32
20	QPR	25	*1*	**QPR**	26
	Totals	1,032	*4*		1,036
	Points Moved		*56*		

the winners against a gray background and the losers in italics. The different aggregate points in the two tables are also shown along with the size of the adjustments to individual team scores.

For the purposes of full disclosure, two of the authors of this book are Liverpool supporters and the one who carried out the *MoTD* analysis is one of them. As can be seen, the 2012–2013 result favors Liverpool, but we honestly do not think this is a result of bias. We cannot be sure, but let us say this: as Liverpool supporters, we hate Manchester United more than any other team, while under our analysis Manchester United gained points in two out of the three years, displacing Manchester City as Premiership champions in 2011–2012. Let us assure readers that we very much did not want this to happen. The team that suffers most under our analysis is Manchester City, losing both their 2011–2012 and 2013–2014 titles under our analysis. But we have nothing against Manchester City and would much rather see them beat Manchester United than vice versa. (Of course, they beat Liverpool to the Premiership in 2013–2014, but the fault lay with Liverpool and Gerrard's slip in the game against Chelsea, whose relentless defensive strategy in that game—a game Liverpool totally dominated in the matter of flair and attractiveness of play even though they could not convert it into goals—has caused us to hate them almost as much as we hate Manchester United.)

So, setting the possibility of bias aside, and bearing in mind that this is not an exact science, if referees' mistakes had been corrected by TV replays in the three seasons in 2011–2014, we show in the tables what would have happened. In sum, in 2011–2012, Manchester United would have won the Premiership instead of Manchester City and Stoke would have been relegated instead of

Table 6.4

2013–2014

Pos	Actual table	Pts	Adj	Adjusted table	Pts
1	Man City	86	–2	Liverpool	86
2	Liverpool	84	2	*Manchester City*	84
3	Chelsea	82	–1	Chelsea	81
4	Arsenal	79	2	Arsenal	81
5	Everton	72	–3	Everton	69
6	Tottenham	69	–8	Tottenham	61
7	Man Utd	64	–4	Man Utd	60
8	Southampton	56	–3	Southampton	53
9	Stoke City	50	–2	Newcastle Utd	49
10	Newcastle Utd	49	0	West Ham Utd	49
11	Crystal Palace	45	–1	Stoke	48
12	Swansea	42	4	Swansea	46
13	West Ham Utd	40	9	Crystal Palace	44
14	Sunderland	38	0	WBA	42
15	Aston Villa	38	–1	Sunderland	38
16	Hull City	37	–5	Aston Villa	37
17	WBA	36	6	Norwich	34
18	Norwich	33	+1	Cardiff	33
19	Fulham	32	–2	*Hull City*	32
20	Cardiff	30	+3	Fulham	30
	Totals	1,062	–5		1,057
	Points moved		59		

Blackburn, Stoke losing a remarkable 10 points. In 2012–2013, Tottenham would have gained the last European Champion's League spot instead of Arsenal. In 2013–2014, Liverpool would have won the Premiership ahead of Manchester City in spite of Gerrard's nightmare against Chelsea, and Hull would have been relegated instead of Norwich.

How the Differences Were Calculated

Tables 6.5–6.7 show how these differences came about.

These tables show, in terms of games, the way points would have been allocated differently in the three seasons according to our analysis. The analysis is separated into "home" and "away." Starting with the left-hand section of the table, a team may gain one, two, or three points or lose one, two, or three points. The rows show in how many games these six possible outcomes happened to each team. The column headed "T" for "Total" corresponds to the "adjusted points" column in tables 6.1–6.3.

The lower rows of the table aggregate the points that would have been gained and lost partitioned into "home" and "away." In 2011–2012, it appears that referees were slightly biased toward teams playing away from home since under our adjustment method they should have given 6 more points in aggregate to home teams and removed 3 total points from teams playing away from home. This is surprising, but the effect is small. In the other two years we see a marked effect favoring home teams, and aggregating over three years the home team bias is also quite marked. In 2012–2013 they should have given 9 points fewer to home teams and 13 points more to away teams; in 2013–2014 the figures are 21 and 16. The three-year aggregate indicates

Table 6.5
Number of games with points won and lost, 2011–2012

		Home						Away						T
		1	2	3	−1	−2	−3	1	2	3	−1	−2	−3	
Man City	89		1					1	1	1	1	2	1	0
Man Utd	89		1					1	2		1			6
Arsenal	70							3		1	1	1	1	0
Tottenham	69		1					1		1	1	1		2
Newcastle	65		1	1							1	1	1	−1
Chelsea	64		1	1	1			2			1		2	−4
Everton	56	2	2		1						1	1	1	−1
Liverpool	52		3			1		1		1		1		6
Fulham	52	1	1		1	1								0
WBA	47	2			1						2		1	−4
Swansea	47	2			1									−1
Norwich	47				1			1			1			−1
Sunderland	45		1			1			1		2	2		−4

		Home						Away						
		1	2	3	-1	-2	-3	1	2	3	-1	-2	-3	T
Stoke City	45	1	1	1	3	2	1		1		1	1	1	-10
Wigan	43		2	1		1				1		1		3
Aston Villa	38		1	1	2				2		3			4
QPR	37		2			1		1					1	0
Bolton	36					2	2		1					-2
Blackburn	31	2				1		1	2		1			4
Wolves	25	1	1	2	2			1	1		1			9
	1047	9	34	15	11	26	15	13	22	15	17	18	18	4
		58			-52			50			-53			
		6						-3						

Table 6.6
Number of games with points won and lost, 2012–2013

		Home						Away						
		1	2	3	−1	−2	−3	1	2	3	−1	−2	−3	T
Man Utd	89	1							1			1		*1*
Man City	78		1			2		1	2		1			*2*
Chelsea	75	1	1		1		1		2					*3*
Arsenal	73					3	1	1		1	1	1		*−6*
Tottenham	72				2	1					1	1		*−5*
Everton	63		2		1			1	2		1	1		*3*
Liverpool	61		2		3				2		2			*3*
WBA	49			1		3		1				1		*−4*
Swansea	46	1	1		1	1		1			1			*0*
West Ham Utd	46	1	1			1			1			1		*1*
Norwich	44	1	1			1		3	3		1	1		*7*
Fulham	43	1	1						1					*4*
Stoke City	42	1	2		2	1		1			1			*1*

		Home						Away						
		1	2	3	-1	-2	-3	1	2	3	-1	-2	-3	T
Southampton	41	1	1					1			2	1		0
Aston Villa	41	1			1						1	1	1	-3
Newcastle Utd	41	1				1			1	1	2			0
Sunderland	39				2	2		1	1			1		-5
Wigan	36	1	1		1		1	2			1		1	-3
Reading	28		1					2				2		4
QPR	25	1			1			1			1	1		1
	1032	11	30	3	15	32	6	15	30	6	15	20	3	4
		44			-53			51			-38			
		-9						13						

Table 6.7
Number of games with points won and lost, 2013–2014

		Home						Away						
		1	2	3	−1	−2	−3	1	2	3	−1	−2	−3	T
Man City	86	1				2		1		1		1		−2
Liverpool	84	1				1		1	1	1		1		2
Chelsea	82			1	1		1	1	1	1	1			−1
Arsenal	79		1	1				1	1		1	1		2
Everton	72				1		1	2	1				1	−3
Tottenham	69					1						3		−8
Man Utd	64	1					1				2			−4
Southampton	56		1			1	1	1				2		−3
Stoke	50	2			1	1	1			1	1	2		−2
Newcastle Utd	49													0
Crystal Palace	45	1			1						1			−1
Swansea	42		2											4
West Ham Utd	40									3				9

	Home						Away						T
	1	2	3	-1	-2	-3	1	2	3	-1	-2	-3	
Sunderland	38					3	1						0
Aston Villa	38			1		1	1				1		-1
Hull City	37	1			1	2	1			1			-5
WBA	36		2					1					6
Norwich	33				2			1	1				1
Fulham	32				1			1					-2
Cardiff	30		1		1			1					3
	1062	9	14	6	5	18	27	8	10	24	7	16	3	-5
		29			-50			42			-26			
		-21						16						

that they should have given 24 points fewer to home teams and 26 points more to away teams, so the home bias appears to be strong. In this it follows the well-analyzed home bias in American sports (see appendix 4).

All these numbers arise out of events that can be described in words, and the tables found in appendix 3 comprise a description of most of the incidents that went into the analysis.

7   Changing the Way Refereeing Decisions Are Made in Football

We know that the football authorities are doing something to improve refereeing decisions by introducing new technologies to make goal-line judgments. In the introduction we provided some figures to explain how little effect this will have on the game. Table 7.1 shows the complete story of the 1,049 matches we analyzed during three seasons of BBC's *Match of the Day* (*MoTD*)—92 percent of all the Premiership matches played in those three years. The table shows the topics of the incidents that the *MoTD* commentators discussed and the commentators' views in respect of whether referees made right or wrong decisions. Once more, this table may not be an exact re-creation of these three seasons of football, but it would have to be literally *incredibly inaccurate* for the overall message we can draw from it to be wrong. That overall message, to repeat, is that goal-line disputes are a near negligible part of the problem of refereeing and the erosion of presumptive justice.

As can be seen, the number of incidents where incorrect decisions about goal-line crossing are said to have been made is very low compared to the many other controversial decisions about which goal-line technology has nothing to say. In the first two seasons, goal-line disputes caused 44 discussions, with referees

Table 7.1
Topic of commentators' discussions during 1,049 matches shown on
Match of the Day: 2011–2014

Officials are:	Right	Wrong	GL Tech	Subtotals	TOTAL
Goal-line dispute	44	5	6		55
Penalty dispute	112	161		273	
Goal-related offside	25	86		111	558
Red card dispute	86	88		174	
Subtotals	267	340	6	558	613

wrong only five times according to the commentators. In the
third season, goal-line technology was introduced, but was
required on only six occasions. In contrast, there were 558 other
discussions about goal-related decisions that could have altered
the course of the match, with referees being wrong 335 times
according to the pundits. According to this analysis, goal-line
technology if employed over all three years could have corrected
at most 3 percent of the total of incorrect decisions. To repeat,
these figures do not have to be exactly correct and the commentators do not have to get everything right for this table to provide an unmistakably strong message. If we halve the numbers
in the middle rows, or even divide by four, goal-line technology
is still providing next to nothing compared to the other categories of goal-affecting judgments.

These other goal-related mistakes could, we claim, be largely
put right using TV replays. But the football authorities say that

this would spoil the game by slowing it down to an unacceptable degree. If TV were used in football in the way it is currently
used to decide "try" or "no-try" in rugby union, or in the case
of run-out or stumping decisions in cricket, they would have a
point. But, as we have argued, there are other ways to use technology in these sports; we will return to them in the conclusion
after we show how to use TV replays in football without causing
disruption.

How could TV replays be used in football where it is considered that even small delays are unacceptable? The answer is
twofold. The first part of the answer is to think about justice, not
accuracy, and this means giving ontological authority to the referee and invoking the Right If Not Wrong (RINOWN) principle
presented in chapter 5. The second part of the answer, unique to
football, is that there should be no interruptions at all; instead,
play should continue while officials viewing TV screens consider
whether a referee has made a glaringly visible mistake. Once
more, unless the mistake is visible enough to be quickly spotted by officials and the crowd, then the referee's decision would
stand and play would continue without interruption. If, however, an obvious mistake is spotted—the kind of mistake that can
be spotted by the commentators on *MoTD*—then play should
be pulled back to where it was when the mistake was made and
restarted in a manner appropriate to the incident.

We want to suggest one more innovation, taken from a principle found in American football. In American football when
referees think that a foul or an unfair move has been made, they
throw a small yellow flag onto the field: commenters say "there
is a flag on the play." The play may continue to its conclusion
before the referees get together, have a discussion, decide what
happened, and, if appropriate, pull play back to where it was,

awarding a penalty of one sort or another. We want to introduce
the idea of "flag on the play" into European football, using not
a flag thrown on the field but a light—a light visible to both the
crowd and TV viewers. The light would be radio-controlled by
the referee, or referee's assistant, or TV judge, any of whom could
choose to have the decision reviewed by the TV judge while play
continued. If no refereeing mistake was spotted upon review, the
TV judge would turn off the light and the game would continue.
If a mistake was spotted, the verdict would be relayed to the
referee who would stop the game and pull it back accordingly.
Using the Right If Not Wrong principle, one would expect the
light never to be illuminated for more than a few seconds.

Table 7.2 shows how the system would work in the case of the
full range of possible infringements and errors. Starting with the
top row, where the referee awards a penalty, the game is halted
anyway and there is time for the TV judge to decide if a mistake
has been made. If a mistake has been made, the penalty decision
will be reversed before the kick is taken and play will be restarted
with a drop-ball or a free kick from the appropriate location. If,
on the other hand, a referee does not award a potential penalty,
then play continues as before, but now the referee, or assistant
referee, might press a button that switches on the light asking
the TV judge to take a second look. The TV judge might decide
to take a second look anyway—he or she would also have that
power to switch on the light. In the table we see "flag" a number
of times. This indicates when the "flag-light" would be turned on.
We will write throughout the remains of this chapter as though
the flag-light idea is acceptable, because it is a convenient way of
describing when an official decides to have a decision reviewed:
they "turn the flag-light on." Of course, the whole system would
still work without any warning light, so the flag-light is not an

essential part of the ideas presented here; but, to us, it seems like a very good idea.

To repeat, as far as what follows is concerned, whoever alerts the TV judge, or the TV judge him- or herself, would, as part of the act of alerting the TV judge, illuminate the flag-light to show that play was being reviewed even as play continued without interruption. If, after a short time—and again, the right length of time would have to worked out as experience accumulated—the TV judge decided the referee's decision was right, play would continue and the flag-light, if used, would be switched off. If the TV judge decided that the referee had made a mistake, and that a penalty should have been awarded, the referee would be informed and would stop the game and award the penalty. Any event that had taken place while the match was under review would be treated as though it had never happened except in the case of foul play or violent conduct, which would still be penalized as appropriate.

Penalty/no-penalty has been dealt with. We now move to the second row, which is about goal-line disputes. This is the kind of case that is now settled by some more complicated form of technology that, although it is often described as "Hawk-Eye," is apparently not a track estimator but a complicated TV replay system. We simply do not know enough about how it works to comment on the use of this technology, but we note that the furor over wrongly awarded goals that led to the introduction of this complex goal-line technology arose out of mistakes that were completely obvious on standard TV replays. Consider the example of the goal scored by England's Frank Lampard against Germany in the 2010 World Cup that was not awarded. It would have taken five seconds for a third official watching TV to have corrected that mistake. There have been a number of similar

Table 7.2
How to use TV replays (TVR) without slowing down the game

Incident	Original Decision	Result of TVR	Extra Delay
Penalty	Awarded	If no penalty then restart with drop-ball, free kick, or goal kick	NONE
	Not awarded	Play on (flag) If penalty then pull back	NONE
Marginal goal	Awarded	If no goal then goal kick	NONE
	Not awarded	Play on (flag) If goal, pull back, restart	NONE
Marginal offside	Flag and play on; do not stop play	If offside, pull back and free kick	NONE
	Not awarded	If no offside, play on	NONE
Clear offside	As now	None	NONE
Foul play, no card	As now	None	NONE
Yellow or red card	Flag and halt play	TVR apprises ref	NONE
Ref misses off-ball incident		(Flag) TVR apprises ref Pull back and free kick, etc.	NONE
Corner throw in		TVR may tell ref to award goal kick or throw to other side	NONE

outrageous errors, but *all of them were easily spotted by standard broadcast TV technology*. It was precisely because they were so easy to see using that technology that they caused such a fuss—TV viewers around the world immediately knew they were being cheated. But by the same token, no higher level of technology was required to correct the mistake and restore the justice of the situation.

Going back to the second row of table 7.2, if this goal-line technology were to be replaced by TV replays, if the referee awarded a goal that was dubious because it was not clear whether the ball had crossed the line, someone could switch on the flag-light, and the third official and the crowd and home viewers could look at the replays and, following the idea of RINOWN, quickly decide to either carry on or overturn the decision. In this case play would already have stopped because a goal would have been awarded. If, in similar circumstances, the on-field referee did not award a goal, the game would continue while he or she or one of the other officials switched on the light; play would continue for the few seconds it would take to decide if the referee had made an obvious error. If no error had been made the light would be switched off and play would continue as though nothing had happened. If a mistake had been made, play would be stopped, a goal awarded, and the normal restart from the center would take place. There would be no delays and there would be no obvious injustice, because there would be no significant difference between the perspective of the viewing public and the match officials.

Moving to the third row of the table, for marginal offside, the game would not be stopped nor would the assistant referee's flag be raised, but the referee or one of the other officials would switch on the flag-light, and—perhaps taking advantage of the

Level 4 technology of computer-manipulated reorientation of the officials' perspective as used by *MoTD* (so long as it is shown to be reasonably accurate)—officials would make a rapid decision. Only if the TV judge concluded that an offside decision was in order would play be pulled back.

Moving to the next row, with clear offside or clear foul play, things would continue as they do now, but with the review taking place during the stoppage created by the on-field decision. Should it turn out that the "clear offside" or "clear foul" was not so clear after all, play would have to be restarted, probably with a free kick. Referees would have to be discouraged from stopping play for these events unless they were really sure they were right.

The situation for yellow and red cards (fifth row) would also be similar to what currently happens. Play is stopped anyway so there would be no extra delay, but during the stoppage the TV judge would confirm the referee's decision or advise against it. Even if the card were withdrawn the free kick would probably still be appropriate. If the "offense" turns out to have been a "dive" or something similar, however, the punishment might be reversed.

Off-the-ball incidents missed by the referee and seen by the TV judge or an assistant referee would lead to their switching on the light, reviewing the incident in public, and if and only if the incident is confirmed, the game would be stopped and the appropriate free kick and or card awarded. We should note—and the point applies to all sports, not just those discussed in this book—that the role of a match official is more than simply spotting and punishing offenses committed by players: the match official also has the duty of keeping the game going and making sure the crowd is entertained. Sometimes it is better to ignore minor fouls so as not to reduce the game to a series of stop-and-start incidents. Wise crowds and commentators generally understand when, for example, a football referee is doing his or her best to avoid reaching for the whistle and preferring to let the game flow. For the same reason, TV-watching officials would be instructed to be careful not to abuse their privilege and not reach for the flag-light switch unless they see something serious in the way of foul play. This, as always, is the kind of human judgment that can never be eliminated from any human decision-making arena, but it is also the kind of judgment that is widely understood.

Moving to the final row, corners and the award of throw-ins could be continually monitored by the third official with the crowd watching the same screen. The game is stopped anyway, and the TV judge would advise if the decision should be changed.

The overarching rule would be that nothing is changed unless there is clear evidence for changing it, where clear evidence means *quick evidence* so the review period would always be short—only a few seconds—and there would be no need to stop the game during the review. As can be seen from the final column of the table, these procedures do not impose extra delays and the game should flow just as it does now. The flag-light would not be designed to be easily visible to players, but teams whose players spotted the light and relaxed their efforts while reviews were taking place would run the risk of being punished by teams who continued to play at full tilt: slowing down during reviews would be a very high-risk strategy.

Whether or not the momentum of goal-line technology means that it will be kept in spite of it being used on only a very small number of occasions, all or most of which could be handled by regular TV replays, we do not know; it is, of course, notoriously hard to understand how the football authorities make their decisions. Table 7.2 shows how football could be restored to the state of perceived justice that obtained before the introduction of television. It would involve no slowing of the game, but it would mean that the public is no longer being palpably cheated by seeing their teams lose games they should have drawn, or won, as a consequence of avoidable refereeing mistakes. It is really hard to see why such a change is resisted.

Conclusion

We have looked at how the science of decision-aid technologies works and have ignored the mythology. Measurement is always subject to error, and decision-aid technologies are not always accurate; they cannot and should not be an infallible replacement for human judgment. There is no exact measurement on the sports field. Accuracy must not become a fetish.

Philosophical thinking can help in the understanding of scientific and technological practice, in this case replacing an obsession with accuracy with a desire for justice for the sports fan. Taking a philosophical look at match-officiating shows how the problem of justice has arisen out of the introduction of TV but can also show how it can be resolved by that same TV technology: broadcast TV created the problem of a mismatch between TV viewers' epistemological privilege and match officials' ontological authority, and broadcast TV can resolve it. Long delays associated with TV occur only when the replays are being used to try to solve an accuracy problem rather than a justice problem. The solution is to restore match officials' ontological authority and accept RINOWN, the Right If Not Wrong principle. Match officials should continue to create what exists in the way of balls, strikes, runs, outs, and not-outs, just as they have always done in the past, unless there is some technology that is already revealing their mistakes to a wide audience. In that case, third officials can use the same technology to put the matter right. But under RINOWN, match officials' decisions are overturned only if the mistake is quickly and glaringly obvious both to the third official and TV viewers. When no mistake is obvious we revert to the presumptive justice that has supported sports throughout history and still supports it at every contest, from a lower-league fixture to a kick-about in the street.

It is true that the decision over whether, for example, a "try" in rugby union has been scored can be interminable when it is supported by TV replays. For those who do not know the game, the decision has to be over whether downward pressure has been applied to a ball that is in contact with the ground beyond the try line. It is like American football in that the ball has to reach the end zone but in rugby it must not just cut the plane of the zone but also be firmly grounded; opposition players may legitimately try to hold the ball up or interpose their hand or bodies between the ball and the turf, and there is often a heap of scrambling bodies around the try line with the ball buried somewhere among them. Officials can take "forever" examining a series of slow-motion replays from a variety of angles, trying to find the optimum view for making what will sometimes be an impossible decision. Such delays are allowed either because rugby union isn't concerned with keeping the game flowing or because rugby union officials do not understand that what they seek should be not accuracy but justice. If they understood this *and* wanted to speed things up, they would return ontological authority to the on-field referee and adopt RINOWN. They would tell on-field referees that in televised games, just like every other game of rugby that has ever been played, their duty is to make a "try/no-try" decision themselves and only refer it to the off-field referee if they have doubts. If the decision is referred, TV replays, visible to the crowd and home viewers, could rapidly show if the on-field referee was clearly wrong, and only in those circumstances would the on-field decision be reversed. It would take seconds, and every TV viewer would see that justice was not being violated. Ever since rugby was invented, referees have been deciding whether a try was fairly scored when the ball is under a heap of struggling bodies without any technological aids. Doubtless many of their decisions were, in some absolute sense, "wrong." But nobody was in a position to make a better decision than the referee (except maybe a player or two at the bottom of the heap, but they can't be allowed to make those kinds of decisions). So, there was no perceived injustice.

Cricket, though it is not a hurried sport, is our model sport as far as the RINOWN principle is concerned, at least when it is a matter of leg before wicket (lbw) decisions. In the lbw case, the umpire makes a decision and then, if there is a challenge, it is up to the third umpire to overturn it. If there is any doubt, the original decision stands. Cricket is a game played as a series of mini-contests with minor intervals between them, and the sometimes drawn-out decisions of the third umpire in the case of a catch, as he or she along with the crowd examines Snicko and Hot Spot, are enjoyable, and all can watch justice unfolding. In the case of run-outs and stumpings, the decision-making process could be improved if the cricket authorities wanted to speed things up. As it is, umpires tend to call immediately for a replay, whereas they could be asked to make a decision first and then only refer it if they have serious doubt. That way the much speedier RINOWN principle could be applied in the case of those decisions too.

In international tennis, there is no waste of time. If a player challenges the umpire's call, the track estimator is called in to adjudicate just as though the umpire had made no judgment in the first place: the entire decision—the entire ontological author-ity—is delegated to the technology (and whoever is sitting in the private control room), and the decision is made quickly with a nice graphic presentation that the crowd can cheer. Everybody is happy—except one or two of the tennis players, who are pretty sure the technology makes mistakes, and, of course, the authors of this book and those who agree with them. We are not happy because we think that presenting virtual reality as real reality is not healthy; we think citizens should understand the world we live in. Once more, we've discussed a range of simple solu-tions. (1) Remove track estimators and bring back TV replays with umpires allowed to view them after they have made a call; once more, if the call was not obviously wrong the original deci-sion should stand. (2) Keep track estimators, but use them, along with the RINOWN principle, to overrule the umpire only when the umpire's call is substantially at variance with the track esti-mator's call. If, according to the track estimator, the ball strikes within, say, 5 mm of the line, the umpire's call should stand. Or, if the crowd has become addicted to the drama, if the ball lands within, say, 5 mm of the line, the track estimator reconstruction should be replaced with a graphic of a spinning coin driven by a random-number generator which will land "heads" or "tails"— "in" or "out." If someone ever investigates the statistics properly, the coin could be appropriately biased.

Paul Hawkins, representing Hawk-Eye and writing in 2015, has claimed that the tennis error is 2.6 mm.[1] This is an improve-ment on past claims, and such an improvement is just what we would expect as the technology gets better. Perhaps the

International Tennis Federation has made more measurements and perhaps they are prepared to make them public. But an average error of 2.2 mm still allows for much larger errors, and we have no idea how these errors are distributed nor how they play out for different kinds of ball impact. To move away from Hawk-Eye in particular, it still seems to us that when a track estimator shows a ball landing close to the line in tennis—say, within the 5 mm range of error allowed by the International Tennis Federation (or maybe, nowadays a millimeter less)—and when the public sees an exact reproduction of a line, a ball, the ball's path, and a reconstructed impact footprint with perfectly defined edges, then the public is seeing *virtual* reality, not real reality. Probably no one knows exactly how closely that virtual reality corresponds to real reality; unless we are missing something—and we are always ready to learn more—a ball shown as a millimeter or so out could in fact be a millimeter in, and vice versa. Viewers are seeing computer-generated graphics, of the kind they see when they watch a movie like *Lord of the Rings*, they are watching what is known in the field of artificial intelligence as a "microworld" (see appendix 4), but they are being allowed to think that the tennis equivalent of Orcs and Elves are real.

Another way of looking at it is akin to those TV series that come under the general title of "Crime Scene Investigation (CSI)" in which the potency of forensic science is mythologized in the way that the accuracy of decision-aid technology is mythologized. In CSI, careful and painstaking investigation will always produce definitive evidence that points without doubt to guilt or innocence. But neither real pioneering science nor real, routine, forensic science are like that—they are full of pitfalls, uncertainties, and, above all, contentious statistical inferences. To make proper use of these devices, if we really want them to make measurements, all this uncertainty needs to be presented. We say "proper" use of these devices, but we have argued that the attempt to reach ever more demanding standards of accuracy is misplaced; the better use for track estimators is in the aid of the construction of archival databases—a new and rather wonderful and rich variant of the history of sports.

The danger of this technology is false transparency: track estimators can present a kind of show-trial justice. And this is not only bad in itself but bad for society. The increasing speed and capacity of computers is making it harder and harder for us to distinguish computer-generated graphics from live video footage; computers, their technology driven by the ever more demanding computer-games industry, are reaching the point where they can generate a lifelike sequence of hue and intensity for every single pixel on a video screen. Once computers can do that in real time, then reality and artificial reality, as represented on the TV screen, will be indistinguishable to the naked eye. Once upon a time dictatorial regimes doctored photographs in the darkroom, removing a figure here, adding one there, so as to produce what looked like realistic evidence to back up their reconstructions of history; nowadays it can happen, or it soon will be able to happen, with video running at full speed in real time. No one can want track estimators to be the thin end of the wedge for the creation of a new kind of gullible public.

But why reach out for devices as complicated as track estimators when, once one understands that the problem is justice, not accuracy, and understands the principle of Right If Not Wrong, TV replays can do most of what we need. The general argument we have tried to mount is that we should always use the lowest level of technology. Consider the cup on the golf green: this could be replaced by a circle to guide the player's aim, while multiple cameras set up around the green reconstruct the path of the ball so that a computer model can determine whether the ball would have fallen into a cup had there been one. Likewise, the bails in cricket could be removed and track estimators could be used to reconstruct path of a ball that brushed or came near to brushing the stumps and reach a conclusion as to whether it had touched or missed. And so on. But why would anyone want to do this? It is obviously much less expensive and much less trouble to work with lower levels of intermediation, and it is also more transparent because there are no unknown error-distributions and everyone can see what is going on. In general, once we eliminate the fetish of accuracy we need very good arguments to move to a higher level of technology.

The philosophical concept of justice versus accuracy and the ideas of ontological authority and epistemological privilege are the spine of the book. They are what enable us to speed up the game with the Right If Not Wrong principle. Things have changed for match officials with the introduction of television. Epistemological privilege increasingly resides with the television-viewing public who tend to have a better view of a disputed decision than the officials judging in real time. We have shown how to put things right.

Afterword

It is two years on since Liverpool beat Manchester City at Anfield in the game described in the introduction. Once more, I, Harry Collins, am in London, watching football with my son; this time we are in a pub. We are both, me especially, still carrying the emotional scars of what happened to Liverpool at the end of 2012–2013. With a brilliant season behind them, having played some of the most breathtaking attacking football that has ever been seen on British television, and with the Premiership nearly in their hands, they lost the trophy to Manchester City when their iconic player and captain, Steven Gerrard, inexplicably slipped and gave a goal away to Chelsea. Then their star player, Luis Suárez, bit another player during the World Cup in Brazil, starting a sequence of events that led to his leaving for Barcelona and Liverpool's form took a dive.

In other ways, nothing has changed. Liverpool is a team whose history is rich with trophies and who spent "umpteen" million pounds on new players before the start of the 2015–2016 season. It is still a team with huge resources, skills, and experience, from one of the large northern towns that have dominated football for decades. Tonight the team is playing in the once-impregnable Liverpool fortress, population half a million, ground capacity 50,000, and tonight, August 17, 2015, the visitors are newly promoted Bournemouth, population 180,000, ground capacity 12,000—nobodies from an effete Southern seaside town mostly known for retirees; Bournemouth, a team that has never played in the top league before. Liverpool is expected to win by a healthy margin: three or four goals. In fact, they win 1–0.

But the truth of the matter is—and, remember, once more, that the writer of this passage is a Liverpool supporter—they should have lost 0–1. In the instant TV replay one could see that the goal they scored was offside, and should have been disallowed, while Bournemouth had scored a goal in the early minutes of the game that was ruled out for foul play. The TV replay showed that the fouled player had exaggerated the offense and the Bournemouth goal should probably have stood. Thus would have occurred one of the major dramas that top-level English football occasionally produces: a team from nowhere beating one of the powerhouses. This time, my son and I are not whooping, because what we have witnessed is Liverpool's continuing decline in form obscured by rank injustice. Yes, we are pleased that Liverpool won, but this is not what we came for. I—and this will probably get me permanently expelled from the fellowship of Liverpool supporters—feel sorry for Bournemouth. The game of football and the Bournemouth fans have been robbed, their hearts have been broken, and the injustice is there for all to see. We hope this book will go a little way to changing things.

Bonus Extra: The Strange Sport of Cricket

Readers of this book from countries where they are unfamiliar with cricket have already had to learn a lot about the game. Having had to put so much effort already, they might want to learn a little more. After all, one website (http://www.topendsports.com/world/lists/popular-sport/fans.htm), using visits to sports websites as a metric, finds that cricket is the second most popular sport in the world; it may be coming your way! The website's ranking of popularity of some top sports with their regional fan bases is shown in table A.1.

This means that learning a bit about cricket might not be a complete waste of time even if you don't live in a cricket-playing-country, and it is, in any case, increasingly possible that you may get a chance to watch it on one of your native TV channels.

Let us start with the bat, the starting position of which in the batter's hands is usually vertical, with the far end—the toe end—tapping the ground. It differs from the baseball bat, having a flat face a little more than four inches wide and twenty-two inches long, held by a round handle at the top about a foot long and an inch and a bit in diameter. It has thick edges about half an inch deep and bulges out at the rear to a depth of about two

Table A.1
The popularity of some top sports according to website usage

Rank	Sport	Fan-base	Regional affiliation
1	Football/ Soccer	3.5 Billion	Europe, Africa, Asia, the Americas
2	Cricket	2.5 Billion	Asia, Australia, UK
4	Tennis	1 Billion	Europe, Asia, the Americas
7	Baseball	500 Million	The Americas, Japan
8	Golf	450 Million	Europe, Asia, the Americas, Canada
9	Basketball	400 Million	The Americas
9	American football	400 Million	Europe, Africa, Asia, the Americas, Australia

inches so as to create a sweet spot eight or nine inches up from the toe end.

The ball reveals a lot more about the game. A cricket ball is a more vicious weapon than a baseball: it is slightly smaller than a baseball, slightly heavier and it is a lot harder. In "test match" cricket, which has traditionally been played during daylight, it is red, whereas in "limited overs" cricket, which is played under floodlights as well as in daylight, it is white. "Test match" and "limited overs" will be explained in due course. The ball has two symmetrical halves held together by a central raised seam, but as the game wears on the players will "work on the ball," shining one side and allowing the other to become rough so as to exaggerate its aerodynamically induced movement through the air. The ball is not exchanged for another unless it becomes very badly damaged or distorted, or hit so far out of the ground that it is lost; wear on the ball is an integral part of the game as it changes its aerodynamic properties and the way it reacts to the bat and bounces off the pitch. If a ball has to be exchanged for another, a similarly worn one will be selected to replace it.

In both baseball and cricket, the ball can be projected by the pitcher/bowler at up to 100 mph with a baseball typically going maybe 5 mph faster than a cricket ball. The reason for the slower speed in cricket is that the bowler must not bend his or her arm but must project the ball with the windmill action of a straight arm. Whereas the pitcher stands still when the ball is thrown, the bowler, who bowls from 22 yards rather than the pitcher's 20 yards, runs in to bowl so as to gather speed and rhythm for the unnatural bowling action.

There are many kinds of bowler, but an initial classification turns on the speed of the ball: fast, medium, or slow. Fast bowlers may run up for twenty or more paces before delivering the ball, whereas slow bowlers may take only a couple of steps. The relative effectiveness of the three kinds of bowler depends on the nature of the pitch and the state of the atmosphere. If conditions are just right, slow bowlers, who impart spin so as to make the ball shoot sideways when it bounces, can be "unplayable." All bowlers can be left- or right-handed, and slow bowlers, otherwise known as "spinners," can bowl different types of ball as their characteristic weapon. Now one must remember the terms "off-side" and "onside," or "leg-side," which were defined in figure 2.2 and chapter 3. A right-handed spin-bowler who releases the ball from the index finger side of the hand will tend to spin the ball clockwise as seen from the bowler's end, and when it bounces it will tend to jag toward the offside of the pitch as defined by a right-handed batter; therefore, such a bowler is known said to bowl "off-spin." "Leg-spinners" bowl out of the little-finger side of their hand and the ball goes the other way. A "googly" is bowled by an off-spinner who, disguising the action, turns the hand over with the fingers pointing back toward the chest so that the ball turns the opposite way from what is expected, while a "chinaman" is a left-hander's googly. Medium-paced bowlers can be very effective when the atmosphere is heavy because they can make the ball "swing" through the air—travel in a sideways curve and also jag off the pitch if the ball is held just so. But some sideways aerodynamic movement and nonballistic changes in trajectory in the vertical plane—"dip," which is mostly the provenance of the spinners—can be employed by all categories of bowler if conditions are right.

Fast bowlers also make the ball jag a little and swing, but one of their chief weapons is intimidation. It is a legitimate part of cricket for a fast bowler to try to hurt or injure a batter by bowling "short" so that the ball rears up and strikes the batter on the hands (gloves are worn but do not give very good protection), the body, or the head. The batter is entitled to try to hit and hurt the fielders who stand very close in the hopes of catching the ball should it loop upward off the bat—fielding positions known, not inappropriately, as "silly mid-off," "silly mid-on," or "silly point." Cricket is not a contact sport, but, given the speed and hardness of the ball and the player's legitimate intentions, it a more dangerous game than baseball. On November 25, 2014, the Australian International batsman, Philip Hughes, was fatally struck by a ball and died two days later; and there is a list of a dozen recognized names going back in time who have died, while injuries are frequent.

To understand cricket it is probably necessary to understand something of social class. As Collins, who has played cricket at a low level, can attest, standing up to a fast bowler requires a lot of courage: one must not back away as the projectile hurtles toward you but must face up to it and allow yourself to be hit if necessary; one must mutter to oneself, "stay in line," "elbow up" (which keeps the bat vertical). Top-class batters sometimes have to try to hit a ball projected at their faces and sometimes the ball gets through and rearranges their features. This kind of courage was what was required of the English upper classes who were expected to lead their men "over the top" from the safety of the trenches and walk calmly into a hail of bullets. And, until fairly recently, cricketers would not wear protective headwear or padding except for pads in front of the lower leg, (since allowing the ball to hit the legs is part of the game); nowadays helmets and various other kinds of padding are worn by the batter and any really close-in fielders, but this has not stopped career-ending or even fatal injuries.

The top-class game also lasts for five days and is played while dressed in pure white tailored clothing—and one can see what that means and guess who wasn't doing the washing and ironing—while the unnatural acts of both bowling and batting are best perfected with long training in top boarding schools—making those initial victories by sides coming from India and the West Indies difficult for some to bear, not to mention the Australians. Batting is unnatural because an "uncultured" swipe at a ball that is bouncing unpredictably will usually result in a miss or a catch, so the ideal stroke is played by keeping the bat vertical with a raised elbow and body behind the ball. The "cover drive," where the ball as struck in this way slightly "inside out," with a slightly bent knee so that it disappears toward two o'clock on the off side, is one the most elegant movements in all sport, with batters sometimes holding the pose for a moment after the ball has disappeared. Collins taught himself to play a good cover drive but, unfortunately, couldn't play any other decent strokes and so did not last long "at the crease." There is a good literature

Figure A.1
Cover drive.

on cricket's notorious early class structure with the amateur gen-
tlemen refusing to share their dressing room with the paid players, and so on. Nowadays, that has mostly disappeared.[1]

But this is only to begin to understand the game; the crucial
point is the role of time and the rhythm of the competition. "Test
matches" are played between national teams and are the classic
form of the game. A test match lasts five days, and its unique feature is that the game is a draw, irrespective of the score, unless all
its phases are completed within the allotted time. Incidentally,
there are three sessions in a day, each lasting two hours. After
the morning session a forty-minute lunch is taken and after
the afternoon session tea is taken—the break lasts for twenty
minutes. In the old days of village cricket (a village match only

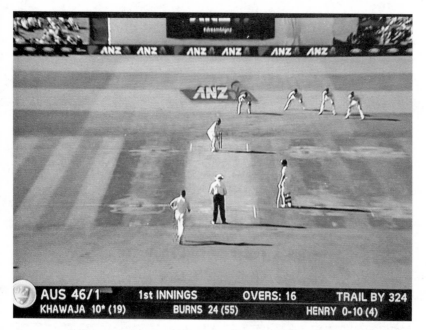

Figure A.2
Wicket keeper and three slips.

lasted a day), the wives would provide a wonderful tea for the players.

The phases that must be completed for a test match not to be a draw comprise two "innings" for each team of eleven players. The British have an "innings" not an "inning," and a completed innings means, not that every player has had a turn "at bat," but that every player has had an innings. You have to know how to use the language: thus the "wicket" is not only the stumps, but is also another term for the pitch. And the pitch is not only the wicket, but also the place on the wicket where the ball bounces: one can say: "That ball was pitched too short," or advise a novice

Figure A.3
Short leg—suicidal.

Figure A.4
What a brilliant catch in the slips!

batter "Watch the pitch of the ball." For those who were follow-ing closely and noticed an inconsistency, since there are eleven players on each team but the game requires that two batters be on the field at once, an innings is complete once ten people are out rather than all eleven so an innings can be completed with one player being "not out" and therefore not having completed his or her innings!

In the case of test matches, rain or bad light can shorten the allotted time making a draw more likely. Cricket, at its best, con-sists of a series of hundreds of violent interactions that are played, overall, at a relaxed pace. Test-match cricket can be deadly dull when batters refuse all risks, and deadly disappointing when bad weather shortens play to such an extent that a tense confron-tation is ruined. But when everything works out just so, it is the most tense and exciting game in the world. Such situations can be stage-managed by a captain who is prepared to surrender some of his or her team's allotted batting rights—cut an innings short before every player has had his or her innings—for the sake of bringing about a result rather than allowing a game to meander to a draw. But "declaring" an innings closed before all the batters are out is fraught with danger because it can offer the other side the chance to win. Thus are created some of the most dramatic moments in sports that can be imagined.

But test-match cricket is too long and too unreliable to bring in enough of the kinds of mass audiences needed for today's sports, so cricket has adapted by inventing various forms of the "limited overs" game. We have seen that there are two batters playing at once with one bowler projecting the ball from one end, and we know that after six balls have been bowled—an "over"—everything switches, including all the fielders, and a different bowler starts attacking from the other end, bowling at whatever batter happens to have wound up there. That is how it happens that if an odd number of runs are scored off the last ball of an over, the batter who hit the ball that results in the runs being scored will find him- or herself still "facing"—in this case facing a different bowler running in from the other end. If there had been no runs or an even number of runs off the last ball of the over, then the other batter would be facing the bowler. And remember that a batter can stay in as long as he or she likes so long as he or she is not out, and there is no need for the batter to try to hit the ball if he or she doesn't want to. In test cricket the primary job of a batter is to last as long as possible without being out, and this imperative has led to some notoriously flawed cricket careers where certain batters would decide that not being out and improving their "batting average" are more important than winning the game for the team. Even though one is less likely to be out if one does not try to score runs, and even though there is nothing in the rules to make you score runs, if runs are not scored at a reasonable pace your team can never win. There have certainly been cases where notoriously slow-scoring batters have been deliberately "run out" by their teammate who is batting at the other end. In test-match cricket, batters have survived for two days and scored 500 runs, though a "century"—100 runs—is what every top batter aims for when they get their turn at the wicket. (Collins never made it to 50 and still remembers the flinch in the face of a fast ball that cost him his wicket the one time he was about to achieve that landmark score!)

Back to limited overs. In the pure form of the game—the test match—the duration is set by the clock, but in limited overs games each side is allowed to bowl a certain number of overs: 50 each in the one-day game, and 20 each in the 20/20 game,

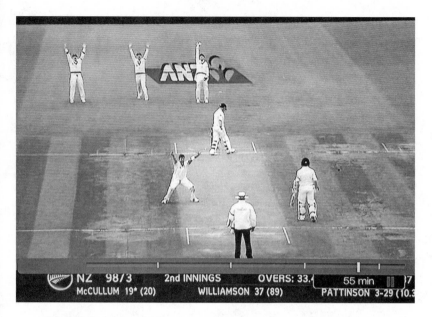

Figure A.5
Howzat?

which is completed in an evening. The game is completed when both sides have bowled their overs. There are no draws, only the remote possibility of a tie if both sides score the same number of runs. The onus on the batting team becomes to score as many runs as possible in the course of the allotted overs. Limited-overs cricket has, therefore, become frenetic and sweaty and, correspondingly, is played in ugly colored uniforms rather than beautiful white clothing.

One way of scoring runs has already been explained—hit the ball and run before anyone can get the ball back from the outfield and break the wicket—but the majority of runs in most matches will be scored by hitting the ball across the "boundary"—a rope

Figure A.6
A run out in a limited-overs game with colored uniforms.

or other marker that defines the outer edge of the playing area. Four runs are scored if the ball crosses the boundary and six runs if the ball crosses the boundary in the air before bouncing: a "six" then, is the equivalent of a home run. In test-match cricket, sixes were a rarity because the aim was to keep the ball on the ground so it could not be caught; this is where a lot of the game's elegance came from—a stroke designed to keep the ball on the ground is a beautiful thing. In limited-overs cricket, however, sixes are important, and the strokes required to hit them are often more like the baseball batter's swipe. Before this becomes too snobby, let it be said that the feedback from limited-overs cricket to test-match cricket, in terms of willingness to take risks and the skills required to execute risky shots without being out, has been greatly beneficial to the longer game and improved the typical test-match as a spectacle.

So long as conditions can be created for the execution of the basic transaction of the bowler and the batters, cricket can take any form. Thus most kids in countries where cricket is popular will play their first games with someone bowling a tennis ball in the street at someone defending pile of coats with a bat cut out of a plank. At the top end, Collins has seen an all-star game played between retired professionals on a baseball field in Texas. Because baseball fields are smaller than cricket grounds, that game too had to be "one-ended," like the street game. But it provided a wonderful display of six-hitting hugely appreciated in a stadium packed with fans—admittedly they seemed to be mostly of Indian descent. But perhaps this very baseball-like version of the game will catch on in in the United States—it is much more action packed than baseball itself.

Appendix 1   A Somewhat Surprising Description of How Hawk-Eye Works

In 2010 Hawk-Eye did produce a more complete description of how it is used at cricket matches.[1] It is operated by three people. Hawk-Eye Innovations describes their roles as follows:

1. Lining up and calibrating the cameras
2. Measuring the pitch and the stumps which do vary from ground to ground
3. 1 member of staff is responsible for the virtual reality graphics and offers LBW replays and all the other Hawk-Eye features to the TV director.
4. The other 2 members of staff both are responsible for the tracking. They work independently of each other to provide redundancy, but are able to see a comparison of the two tracks. If they are different for any reason, they can be pro-active in working out why rather than being re-active after a LBW appeal. On a ball by ball basis they would do the following:

(a) hit a button to tell the system that a ball has been bowled and trigger the tracking

(b) manually fine tune the point on the trajectory where interception with the batsman was made. Automatically the system is only able to determine the interception point to the nearest frame of Hawk-Eye video running at 106 frames per second. This can be improved manually and is the only way to ensure that the interception point is accurate to 5 mm.

(c) Tune settings to account for varying light conditions.

(d) Tune settings to deal with camera wobble.

We learn from this that at this time the frame rate was around 100 fps. But what is really interesting is point (b)—the need for manual intervention to maximize accuracy. The remarks about the need for various kinds of fine-tuning are also revealing. This description intimates that the Hawk-Eye system is far less autonomous than we had been given to believe and that there could be scope for mistakes driven by human error. This reinforces the need for a more transparent presentation of how these devices work and, especially, the extent to which there was or was not manual intervention in their early history.

Appendix 2   An Early Technical Description of Hawk-Eye

We did manage to obtain a technical paper that appears to have been written by the original developers of Hawk-Eye. The paper is N. Owens, C. Harris, and C. Stennett, "Hawk-Eye Tennis System," in *Visual International Conference on Information Engineering 2003 (VIE 2003)* (London: Institution of Electrical Engineers), 182–185. This technical article was difficult to access in two senses: it was written in esoteric language; and we could not download the article from home without paying a fee or joining a professional organization, nor could we find the authors' homepages to ask them directly for a copy. It was possible to discuss the paper because we found we could download it from a Cardiff University–based computer.

Since, whether we like it or not, this book, where it deals with track estimators, is largely based on material in the public domain, we thought it worthwhile to reserve the major discussion of this source for an appendix so that our main arguments about track estimators can be used as an example of what an assiduous researcher can do using only readily available sources. In this way it is a pale echo of Theodore Postol's work in investigating the claims made about the effectiveness of the Patriot Missile in shooting down Iraqi Scud rockets during the first Gulf

War.[1] Incidentally, as of February 2016 Google Scholar reports that the Owens, Harris, and Stennett article has been cited in 53 other articles. We checked a number of these that could be downloaded from home but nothing we found bore on the questions discussed in this book. There were discussions of how to track a ball using a single camera, discussions of better kinds of software, and so on. Those we could not download gave no indication in their titles that they would bear on the concerns of this book.

The original developers of Hawk-Eye appear to be "Roke Manor Research," a private engineering consultancy—which seems to have developed the tennis-line-calling system on behalf of Hawk-Eye Innovations. The Owens et al. paper acknowledges that Hawk-Eye Innovations paid for the research. Crucially, and perhaps surprisingly, the paper does not consider in any detail the questions concerning accuracy analyzed above. Rather, it explains the broad principles of the system. The paper is extremely densely written and includes some technical terms that only a specialist would understand. Nevertheless, the general drift can be grasped after a couple of readings if one has had some experience with technical literature. That said, what follows is written under the standing caveat that the authors of this book are not engineers and would, in the normal way of our research, have taken this paper as a basis for discussion with its authors, or similarly qualified persons, rather than as something that we would try to interpret entirely on our own.

We believe we can learn from the paper that one of the things we believed would be useful—showing data points on the reconstructed image of the flight path of the ball (Question 7)—is impossible. We thought that if the path showed the position of each individual frame from which the path was estimated, it would help the viewer judge the accuracy or the reconstruction—the denser the points, the better the reconstruction. But we learn from this paper that each flight path is built up from what are called "tracklets." which are short two-dimensional sectors of the path between TV images. The tracklets produced by the three cameras are built up into a three-dimensional image quite late in the reconstruction process, so there may be no equivalent of what we call "data points"—only an unfiltered jumble of data point "candidates" pertaining to individual cameras. This, however, does not affect the principle of our arguments; in the last resort, the apparatus does have to build its model of the trajectory from data gathered from a series of camera frames, so the data point idea works even if these data are assembled in a more complex way than our analysis intimates. What is very, and somewhat surprisingly, unclear in the technical paper is how the bounce point is determined and how its accuracy might be affected by frame rate.

We believe the paper shows that it is worth asking questions about the accuracy with which the position of the ball is located in any one frame. The paper (written in 2003) suggests that Hawk-Eye uses output from the existing broadcast TV cameras rather than utilizing special cameras of its own with higher frame rates (Question 2). The paper explains that cameras have to be calibrated and the "observed" ball position at any one time has to take account of pan, tilt, and zoom. The paper seems to say that the position of the ball has to be interpolated from the position of the lines on the tennis court and that these are sometimes obscured or out of frame; it seems to say that these lines are themselves reconstructed from "virtual line centers." In other words, there is at least some room for error even in constructing the model of the court from which the location of the ball in any one frame is estimated. It does say that, depending on the shutter speed of the camera and the speed of the ball, the ball may sometimes appear as a circle and sometimes a "sausage-shaped streak." The paper says that in a long shot the ball may be represented on the image by as few as two pixels.

The paper does seem to illustrate the difficulty of reconstructing short trajectories, in this case in reference to the half volley in tennis. Thus, a passage on the fourth page of the paper reads:

Special code was written to deal with the case of a half volley. Often the upward bouncing portion of the half volley was too short to be extracted, and so an estimated "dummy" track was inserted in the impact point chain. The bounce point was determined using a one-sided Kalman Filter applied to the incoming track. Configurable bounce retardation percentages were used to estimate the initial velocity of the upward bouncing track at the instant just after the bounce. This was used to create a purely ballistic intermediate track, with an acceleration a of [0, 0, –g]T. This dummy track was inserted in the chain of tracks, providing an unbroken set of tracks for the visualisation module.

We understand this to mean (under the standing caveat that we are not engineers), that the upward portion of bounce in the case a half volley often did not allow the collection of usable data. It was therefore estimated to be a function of the velocity prior to bounce, the loss of speed due to the bounce, and the deceleration due to gravity.

The paper appears to imply that in tennis both prebounce and postbounce trajectories or "tracks" are used in the estimation of bounce point.

We understand the paper to reveal that the court lines are continually reconstructed in real time. The lines are photographed, a virtual center is constructed from the photograph for each line in short sections, and these are then joined to reproduce the straight center of the lines as a mathematical object with zero width. The position of the edge of the line is then reconstructed, we guess, by adding a nominal half-line width to this virtual central line. This would seem to suggest another possible source of difference between the virtual world reconstructed by a track estimator and the real world. It is worth noting that all these uncertainties could be resolved in a short conversation or a few emails.

The very density of the description of how the position of the ball and the bounce was estimated, and how the track seen by the viewer was reconstructed, seems to reinforce the argument we made above that the only sure test of the accuracy of a device such as Hawk-Eye is direct measurement of errors in comparison to some more absolute test of position such as a very high-speed camera.

Appendix 3   Description of the 335 Football Refereeing Mistakes Listed in Table 7.1

Table A3.1

Wrong refereeing decisions: Season 2011–2012

Date	Score	Referee	The Play	Commentators' Discussion
Aug. 13, 2011	Liverpool 1 vs. Sunderland 1	Phil Dowd	Penalty to Liverpool as Richardson brings down Suárez. Looked like a foul in real time and replay confirms it was the right decision. Liverpool want him sent off for denying a goal-scoring opportunity, but he only gets a booking. I think the referee gets it right with the penalty, but Richardson should have been sent off for denying a goal-scoring opportunity.	Alan Hansen says it's no doubt there's a penalty kick. Hansen thinks it's nonsense; Richardson should have been sent off.

Table A3.1 (continued)

Date	Score	Referee	The Play	Commentators' Discussion
			Carroll scores for Liverpool, but it's chopped off for a foul on Ferdinand. Looked soft in real time. Replay shows there was small push in the back, but I don't think there was much in it. Decision could have gone either way.	Hansen thinks there's minimal contact, and on another day that goal stands.
Aug. 13, 2011	Newcastle Utd 0 vs. Arsenal 0	Peter Walton	Joey Barton accuses Songe of stamping on him. I didn't see it at the time, and neither did the referee. Replay shows they go in for a challenge, Songe gets up and deliberately stamps on his ankle. Should have been a red card.	Hansen thinks this is a total lack of discipline. It's malicious, it's nasty. He thinks it's a certainty the Football Association will be in touch.
Aug. 14, 2011	Stoke City 0 vs. Chelsea 0	Mark Halsey	Chelsea appeal for a penalty, as Shawcross hacks down Torres. My first thought was that it was a penalty. Replay shows it was a foul and it was inside the box.	Lee Dixon thinks this should have been a penalty.
Aug. 20, 2011	Arsenal 0 vs. Liverpool 2	Martin Atkinson	Liverpool score. Replay shows Suárez was offside in the build up to the own-goal.	Hansen thinks he's probably offside.

Table A3.1 (continued)

Date	Score	Referee	The Play	Commentators' Discussion
Aug. 20, 2011	Sunderland 0 vs. Newcastle Utd 1	Howard Webb	Newcastle appeal for hand-ball on the goal line by Larsson. Was hard to see in real time. After consulting the assistant referee, the referee gives corner. After watching the replay it's a deliberate hand-ball. Should have been a penalty to Newcastle and Larsson should have been sent off.	Mark Lawrenson shows the poor view Howard Webb has. Should have been a penalty and a red card.
Aug. 21, 2011	Norwich City 1 vs. Stoke City 1	Neil Swarbrick	Stoke's Walters is awarded a penalty after Leon Barnett bundles him over in the box. Barnett is sent off for denying a goal-scoring opportunity. Looked like a clumsy challenge and the right decision in real time. Replay shows challenge was actually outside the box. Poor decision to give penalty, although it's still a red card.	John Hartson thinks it's definitely not a penalty. Assistant referee has to help out here. It's a coming together of players, but by letter of law it's a sending off. Dixon doesn't even think it's a foul.
Aug. 27, 2011	Blackburn 0 vs. Everton 1	Lee Mason	Everton get a penalty as Fellaini is fouled by Samba. Looked soft in real time I thought. Samba was climbing on his back, but you see twenty incidents like that every game.	Alan Shearer can't believe that decision. It's never a penalty. He notes that the referee took his time, because he doesn't think he's sure about the decision. Dixon doesn't think this is a penalty either.

Table A3.1 (continued)

Date	Score	Referee	The Play	Commentators' Discussion
Aug. 27, 2011	Aston Villa 0 vs. Wolverhampton Wanderers 0	Martin Atkinson	Wolves player is barged down in the box after a collision in the air. I think it's a penalty as Herd hits Johnson well after the ball has gone. Has to be a penalty.	Dixon doesn't think it was a penalty. He's not looking at Ward. Shearer says that could have very easily been a penalty. Given that the pundits are split I will go with my own judgment, which is that it was a penalty.
Aug. 28, 2011	West Bromwich Albion 0 vs. Stoke City 1	Mike Dean	Shotton scores for Stoke as Ben Foster appeals that the ball was kicked out of his hands. Hard to tell in real time. Replay shows the keeper was in possession of the ball just before Shotton kicks it out his hands. Goal shouldn't have stood.	Dixon thinks it's a foul. Roberto Martinez thinks it's in the keeper's hands and therefore it's a foul. It's down to the referee's angle, ultimately. Hansen says foot is up.
Sept. 10, 2011	Stoke City 1 vs. Liverpool 0	Mark Clattenburg	Liverpool appeal for hand-ball as it strikes the hand of Rory Delap. Looked like a penalty in real time as his hand was out and high. Replay confirms it should have been given.	Hansen thinks this is a stonewall penalty kick.

Table A3.1 (continued)

Date	Score	Referee	The Play	Commentators' Discussion
			Stoke wins penalty as Carragher brings down Walters. Looked soft in real time. Replay shows it was a very soft decision; I think Walters has thrown himself down, but Carragher does have his arm around his waist so I can see why the referee thinks he's impeded him.	Hansen doesn't think there is a lot of contact, and he's gone down very easily. He doesn't think that's a penalty kick.
Sept. 10, 2011	Everton 2 vs. Aston Villa 2	Michael Oliver	Leighton Baines appeals for a penalty as he is brought down by Bannon. Looked like a clear penalty at the time. Michael Oliver had a great view, but gave nothing. Replay showed it was a clear penalty. Poor decision.	Hansen thinks this is a penalty kick every day of the week. How he doesn't give it is beyond him. Gary Lineker says "He is so close." Shearer says it's a stonewall penalty.
Sept. 11, 2011	Norwich City 0 vs. West Bromwich Albion 1	Mark Halsey	Penalty to West Bromwich Albion as Steve Reid goes down in the box. Looked soft in real time. Replay shows it was soft. Reid went down far too easily. Poor decision.	Dixon feels the penalty was very harsh. He thinks Reid goes to ground very softly. Brad Freidel thinks he's playing for it there.

Table A3.1 (continued)

Date	Score	Referee	The Play	Commentators' Discussion
			West Bromwich Albion's Tamas appears to elbow Norwich City player in the box, amidst appeals for a penalty. Hard to tell in real time. Replay shows the West Bromwich Albion defender did swing an elbow which catches Vaughan and bloodies him. Should have been a penalty.	Dixon didn't see it in real time. If referee had a good view of it, then he would have given him a red card and given Norwich City a penalty. Freidel says from Halsey's view he mustn't have seen it. Otherwise he would have given a penalty.
Sept. 17, 2011	Blackburn 4 vs. Arsenal 3	Andre Marriner	Yakubu scores for Blackburn. Looked offside in real time, but goal is given. Replay shows he was marginally offside.	Shearer thinks he's probably just offside.
Sept. 18, 2011	Manchester Utd 3 vs. Chelsea 1	Phil Dowd	Chris Smalling scores for Utd. Replay shows he was marginally offside. Poor decision by the assistant referee.	Hansen thinks this is offside.
Sept. 18, 2011	Fulham 2 vs. Manchester City 2	Mark Clattenburg	Manchester City appeal for free kick in the build-up to Fulham's equalizer. Looked like a foul at the time, as Džeko was caught by Kelly.	Hansen thinks this definitely a free kick. Poor decision.

Table A3.1 (continued)

Date	Score	Referee	The Play	Commentators' Discussion
Sept. 24, 2011	Stoke City 1 vs. Manchester Utd 1	Peter Walton	Hernandez is brought down in the box and Utd appeals for a penalty. Looked like a stonewall penalty at the time. Replay shows Woodgate caught him. I think it should have been a penalty. Conceivably Woodgate could have been sent off as well for denying a goal-scoring opportunity.	Lawrenson thinks it should have been a penalty and a red card. Doesn't matter what time of the game it is; a penalty is a penalty. Lineker also thought it should have been a penalty and a red card. Woodgate pushes him in the back. Hansen says it's a penalty kick, red card, and the result would have been absolutely different.
Sept. 24, 2011	Liverpool 2 vs. Wolverhampton Wanderers 1	Kevin Friend	Wolves appeal there is a foul on Johnson by Andy Carroll in the build-up to the first goal. Hard to see anything in real time. Replay shows Carroll did push him and I think it should have been a foul.	Hansen thinks it's a push. Lineker says it looks like a bit of a push.

Table A3.1 (continued)

Date	Score	Referee	The Play	Commentators' Discussion
Sept. 25, 2011	Queens Park Rangers 1 vs. Aston Villa 1	Michael Oliver	QPR appeal for hand-ball by Alan Hutton. Hard to see in real time. Replay shows it did clearly hit Hutton's hand. Should have been a penalty.	Robbie Savage thinks this was a stonewall penalty. There are eight yards between ball and arm and he moves arm toward it. Shearer says distance between the ball and hand is why it is a penalty.
Oct. 1, 2011	Wolverhampton Wanderers 1 vs. Newcastle Utd 2	Mark Halsey	O'Hara is brought down and Wolves appeal for a penalty. Mark Halsey points and says it was outside the box, so gives a free kick. It looked in the box I thought, and replay confirms the infringement was inside the box. Poor decision; it should have been a penalty.	Lineker says it was in the box. Shearer thinks it's a clear penalty, and he called it straight away. He's two yards inside the box.
			Doyle scores for Wolves, but it's chopped off because linesman said the cross went out of play before being headed back into the goal-mouth. Looked tight at the time. After watching the replay it's very hard to tell whether it went out or not, but I don't think it did.	Dixon says it doesn't look out to him. He doesn't think the linesman is in position and is therefore guessing. Goal should have stood.

Table A3.1 (continued)

Date	Score	Referee	The Play	Commentators' Discussion
Oct. 1, 2011	Everton 0 vs. Liverpool 2	Martin Atkinson	Jack Rodwell is sent off for Everton. I didn't think much of it at the time. After watching the replay, I thought it was maybe a booking at worst. Strange decision by the referee. That was never a red card.	Dixon says the referee was so inconsistent today. He says if that's a red card then it's going to become a noncontact sport. He goes for the ball and his foot is not high. Lineker says it was incredibly Harsh. Shearer says there's no way that's a red card. Referee has to look after the fact and say he's wrong.
Oct. 1, 2011	Aston Villa 2 vs. Wigan 0	Mark Clattenburg	De Santo goes down and appeals for a penalty to Wigan. I thought it was a foul in real time and replay shows Collins caught him. Poor decision.	Dixon thinks it was a penalty. Game might have been different if it was given.
Oct. 2, 2011	Tottenham 2 vs. Arsenal 1	Mike Dean	Van der Vaart scores for Tottenham. After several replays commentator raises the point that he might have controlled it with his arm rather than his chest before volleying into the corner. Was very difficult to tell after watching the replay several times. There was no appeal from any Arsenal players.	Dixon thought at the time it was hand-ball. It's very difficult for the officials to see. He doesn't think the officials can see it. Dion Dublin thinks it's a hand-ball.

Table A3.1 (continued)

Date	Score	Referee	The Play	Commentators' Discussion
Oct. 2, 2011	Bolton 1 vs. Chelsea 5	Peter Walton	Bolton appeals ball went over line as Ivanović cleared for Chelsea. I thought the ball was over the line at the time of the incident, as Ivanović was in the net. Replay shows the ball was over the line. Poor decision by the assistant referee.	Pundits didn't discuss, but I included it in my analysis. The decision didn't affect the game as Chelsea was up 5–1 at the time.
Oct. 2, 2011	Swansea 2 vs. Stoke City 0	Mike Jones	Wilkinson goes in with a shocking challenge on Swansea's Dyer. Replay shows it was a really bad challenge. He should have been sent off.	Colin Murray says that's a Bruce Lee challenge.
Oct. 15, 2011	Liverpool 1 vs. Manchester Utd 1	Andre Marriner	Ferdinand catches Adam and he should have received a second yellow card. Replay shows there was contact. Adam was driving into the box and I think that's a booking.	Hansen says Rio Ferdinand should have been sent off there.

Table A3.1 (continued)

Date	Score	Referee	The Play	Commentators' Discussion
Oct. 22, 2011	Aston Villa 1 vs. West Bromwich Albion 2	Phil Dowd	Penalty to West Bromwich Albion as the assistant referee gives a foul for an off-the-ball infringement by Herd against Olsson. Herd is sent off for the offense also. Hard to tell in real time, but replay shows that Herd appears to make a stamping motion on Olsson off-the-ball in the box. Olsson gets straight up so there doesn't look to be much in it. Olsson was grabbing his leg at the time, which provoked the reaction from Herd.	Hansen said he can't see why on earth the assistant referee has sent him off. Lawrenson agrees and says the player is trying to get his leg out. He says even Olsson doesn't react. He needs to be 100% certain that he's stamped on him, and I don't think he can be sure.
Oct. 23, 2011	Manchester Utd 1 vs. Manchester City 6	Mark Clattenburg	City appeals for penalty, as Richards goes down in the box. I thought it was a penalty at the time, and replay confirms Anderson took Richards out. Poor decision.	Hansen thinks this should have been a penalty kick.

Table A3.1 (continued)

Date	Score	Referee	The Play	Commentators' Discussion
Oct. 23, 2011	Queens Park Rangers 1 vs. Chelsea 0	Chris Foy	Appeals for penalty by Chelsea players as Lampard goes down under challenge from Fitz Hall. I thought it was a penalty in real time. Replay confirms he was dragged back by defender. Poor decision.	Pundits didn't discuss. QPR manager Neil Warnock admitted this should have been a penalty against his team. On the basis that the opposition manager thought that this should have been a penalty against his team, I've decided to include this incident within my analysis.
Oct. 29, 2011	Chelsea 3 vs. Arsenal 5	Andre Marriner	Arsenal's goalkeeper Szczęsny takes out Ashley Cole and is booked. Chelsea players appeal for red card. I felt it was a stupid challenge from the keeper in real time and that he should have been sent off. Replay shows keeper did foul him, and I don't think the defender will stop him putting it in the net. Poor decision.	Hansen feels the keeper should have been sent off. Shearer feels he has denied a goal-scoring opportunity and should have been sent off.

Table A3.1 (continued)

Date	Score	Referee	The Play	Commentators' Discussion
Oct. 29, 2011	Norwich City 3 vs. Blackburn 3	Anthony Taylor	Norwich City get penalty in stoppage time, as ball comes off N'Zonzi's hand. I felt it was harsh on Blackburn after watching the replay. It struck his hand without him knowing much about it at close range. On top of that, I thought he was being fouled. Poor decision.	Shearer feels they are very unlucky. Shearer thinks he's being fouled, and to give that as a hand-ball is very harsh. Lineker said it's a ludicrous decision.
Nov. 6, 2011	Fulham 1 vs. Tottenham 3	Peter Walton	Loud appeals from Fulham for a penalty kick during a goal-mouth scramble. Replay shows that Kyle Walker did handle it on the ground and that it should have been a penalty to Fulham.	Dixon thinks that Fulham should have had a penalty here. Fulham can count themselves unlucky.
Nov. 6, 2011	Wolves 3 vs. Wigan 1	Lee Probert	After a Wigan corner, there is an altercation between several players and Wigan's Alcaraz appears to spit at a Wolves player (Steerman). He should have been sent off.	Dixon thinks it's the worst thing that can happen to you as a player. Horrible thing to do.
Nov. 19, 2011	Wigan 3 vs. Blackburn 3	Andre Marriner	Victor Moses goes down in the Blackburn box. Looks like a penalty in real time. Replay seems to suggest Moses was brought down also. Poor decision.	Hansen thinks this was a stonewall penalty. Referee was in a perfect position, and yet he doesn't give it.

Table A3.1 (continued)

Date	Score	Referee	The Play	Commentators' Discussion
			Bizarre incident. Yakubu wins corner for Blackburn, grabs ball and places it down on corner circle. Pederson then runs over and starts dribbling toward goal, before laying it off for a Blackburn goal. I assume the referee thinks Yakubu touched it to him, but replay shows he placed it on the ground. Officials say after the game that Yakubu touched it.	Hansen thinks Yakubu didn't hit it at all. If he did hit it, the referee wouldn't have turned his back. Hansen highlights Pederson and Yakubu talking to each other and laughing about it. Lineker thinks it was cheating.
Nov. 19, 2011	Stoke 2 vs. Queens Park Rangers 3	Mike Jones	QPR's Joey Barton brings down Jermaine Pennant. It looks soft. Replay shows Barton stuck out his leg and there was minimal contact, but I don't think it is a penalty.	Shearer thinks it was a poor decision from Mike Jones. The referee is only two yards away. Hansen agrees that it wasn't a penalty.
			Joey Barton brings down Stoke's Robert Huth, but referee waves play on. It looked soft in real time. Replay shows, however, that Barton caught him and didn't get anything on the ball. Definite penalty.	Hansen thinks it was stonewall penalty. Shearer agrees it was a penalty.

Table A3.1 (continued)

Date	Score	Referee	The Play	Commentators' Discussion
Nov. 26, 2011	Manchester Utd 1 vs. Newcastle 1	Mike Jones	Newcastle's Ben Arfa is brought down by Rio Ferdinand in the penalty area. Referee gives penalty. Looks touch-and-go in real time, although strong reaction from Manchester Utd's players that it wasn't. Replay confirms that Ferdinand played the ball. Poor decision.	Lineker thinks it was blatantly a bad one. They get the decision horribly wrong, according to Shearer. The referee has made the fatal mistake of letting the linesman change his mind. How on earth he can't see that Ferdinand has got the ball is beyond him. Terrible decision.
Nov. 26, 2011	Stoke 3 vs. Blackburn 1	Mark Halsey	Blackburn's N'Zonzi elbows player in the box. Should have been a red card and a penalty. Referee missed the incident.	Lineker suggests N'Zonzi might be in a bit of trouble. Shearer agrees and thinks he might be getting a call from the FA.
Nov. 27, 2011	Liverpool 1 vs. Manchester City 1	Martin Atkinson	City's Mario Balotelli is sent off for second yellow card. First yellow was a booking. I think the second yellow is a little soft. Harsh decision to send him off.	Lawrenson says he does daft things. Lawrenson thinks this is a bit of nothing. Michael Owen also thinks there's not much in this.

Table A3.1 (continued)

Date	Score	Referee	The Play	Commentators' Discussion
Nov. 27, 2011	Swansea 0 vs. Aston Villa 0	Neil Swarbrick	Villa appeals for penalty as ball comes off the hand of Williams. Looked like a penalty at the time and replay shows it was a clear hand-ball. Poor decision.	Pundits didn't discuss. I have decided to include this incident, despite the fact the pundits didn't discuss the incident. I do so because the pundits were discussing Gary Speed's death, rather than incidents. The penalty was blatant though and should have been given.
Dec. 3, 2011	Newcastle 0 vs. Chelsea 3	Mike Dean	David Luiz of Chelsea brings down Ba. He looks to be the last man. Free kick is given, but only a yellow card shown. Newcastle are incensed by the decision. Replay confirms it is a definite foul and he needs to be red-carded as he is denying a goal-scoring opportunity. Poor decision.	Hansen thinks he should go. Luiz is in a bad position. Hansen thinks the referee thinks the ball is going to go through to the keeper, but he disagrees. To him, it's a certain red card. If he goes the result is different.

Table A3.1 (continued)

Date	Score	Referee	The Play	Commentators' Discussion
Dec. 3, 2011	Tottenham 3 vs. Bolton 0	Stuart Attwell	Bolton's Gary Cahill is sent off for bringing down Scott Parker. Referee has deemed that he was last man. Replay confirms this is very harsh as they were just over halfway line and there was a Bolton defender near the middle of the park in line with them.	Hansen thinks it could have gone either way, but it looked harsh. He notes that the referee is 26 yards away. Spurs manager Harry Redknapp thinks the decision was a bit harsh. He felt for them. It looked harsh, I haven't seen it again, but couldn't believe it when red card came out.
Dec. 3, 2011	Queens Park Rangers 1 vs. West Bromwich Albion 1	Martin Atkinson	Shaun Wright-Phillips scored for QPR but it was chopped off for offside. Looked off in real time. Replay confirms that he was onside and goal should have stood. Poor decision.	Hansen says you're 1–0 up, you're pushing for a second goal, so that really hurts. Great goal and Warnock is quite rightly incensed. Bad decision.
Dec. 4, 2011	Wolverhampton Wanderers 2 vs. Sunderland 1	Phil Dowd	Sunderland get a penalty as Larsson is brought down by Craddock. Looked like a stonewall penalty in real time as defender stuck his leg out. After watching the replay, it looks like Larsson dived over the outstretched leg, and the challenge was marginally outside the box. Poor decision.	Savage says it should never be a penalty. He shouldn't stick his leg out, but it's outside and he's not been touched. Pat Nevin says he's let himself down because that's a dive with bells and whistles on.

Table A3.1 (continued)

Date	Score	Referee	The Play	Commentators' Discussion
Dec. 10, 2011	West Bromwich Albion 1 vs. Wigan 2	Mike Dean	West Bromwich Albion forward goes down in the box amidst big appeals for a penalty. Referee books him for diving. I thought he was fouled at first watch; after watching the replay I think the referee got the decision wrong to not give a penalty. There was contact.	Hansen thinks it's a penalty kick. He's just clipped and there's enough contact. Definite penalty kick. Shearer says it's a penalty, and it's unfortunate for West Bromwich Albion.
Dec. 11, 2011	Stoke City 2 vs. Tottenham 1	Chris Foy	Tottenham appeal for hand-ball by Peter Crouch, as he sets up Etherington for a goal. Was hard to see in real time. Replay confirms it was a clear and deliberate hand-ball.	Dixon says Chris Foy made some really big mistakes. He says this is hand-ball, and can't believe that the assistant referee hasn't seen it. David O'Leary thinks the referee was in a good position to see this too.
			Spurs appeal for penalty as Kaboul is pulled back from a corner. Replay shows there was a clear pull and it should have been a penalty to Spurs. Poor decision.	Dixon says this is a clear penalty.

Table A3.1 (continued)

Date	Score	Referee	The Play	Commentators' Discussion
			Spurs appeal for penalty as Shawcross handles on the goal-line. Hard to see in real time. Replay shows it was a poor decision as he clearly handled. Should have been a penalty and a red card for Shawcross.	Dixon shows that this is a clear hand-ball by Shawcross.
			Adebayor scores for Tottenham, but it's chopped off for offside. He looked off at the time, but the replay shows he was well onside. Really poor decision.	Dixon can't believe this is given as offside.
			Kaboul sent off for Spurs as he brings down Walters. Second yellow was justified. His first yellow was for dissent, after he argued about the penalty that wasn't given.	Dixon doesn't even think the second yellow card was a booking. O'Leary says the referee has to show common sense with such decisions, otherwise there'll be no one on the field.
Dec. 11, 2011	Sunderland 2 vs. Blackburn 1	Peter Walton	Blackburn scores, but it's chopped off for foul on keeper by Christopher Samba. I think it's a poor decision to give a free kick there. The keeper misjudged the flight of the ball and he got away with that.	O'Leary thinks this should have been a goal. Dixon also thinks this was a legitimate goal.

Table A3.1 (continued)

Date	Score	Referee	The Play	Commentators' Discussion
Dec. 17, 2011	Wolves 1 vs. Stoke 2	Anthony Taylor	Stoke's Jonathan Woodgate brings down Matt Jarvis. Looks like a penalty in real time. Replay confirms Woodgate catches him and misses the ball. Good decision. However, Woodgate had already been booked and thus should have received a second yellow.	Hansen thinks Woodgate should have been sent off. Certain penalty and should have received second yellow.
Dec. 17, 2011	Everton 1 vs. Norwich City 1	Lee Probert	Norwich City's Grant Holt and Everton's Fellaini are tussling. Holt elbows Fellaini and is lucky to get away with not being sent off.	Lawrenson says Holt catches Fellaini, but it's like the pot calling the kettle black. Lawrenson did imply that he should have been sent off, however.
Dec. 18, 2011	Manchester City 1 vs. Arsenal 0	Phil Dowd	Van Persie scores for Arsenal, but it's chopped off for offside. Looked tight in real time. Replay shows he looked level. The assistant referee should favor the attacker.	Dixon feels this was a split decision. He thinks he's level. Difficult for the linesman, but it's decisions like this that change big games.
			Arsenal appeal for penalty, as Micah Richards handles in the box. I thought this was a clear penalty. His arms were out, and there was movement toward the ball. Poor decision.	Dixon thinks this was a clear penalty, and can't understand why referee hasn't given it.

Table A3.1 (continued)

Date	Score	Referee	The Play	Commentators' Discussion
Dec. 21, 2011	Aston Villa 1 vs. Arsenal 2	Jon Moss	Van Persie goes down in the Villa penalty area and is booked for diving. I thought there was contact and it should have been a penalty. Poor decision.	Lineker thinks it was a penalty and a crazy booking. Says it should be rescinded.
Dec. 26, 2011	Manchester Utd 5 vs. Wigan 0	Phil Dowd	Wigan's Sammon is sent off for foul on Carrick. Carrick was clutching his face after incident, but in real time it was difficult to see how bad it was. Replay showed it was incredibly harsh. He used his arm to shield him off, and he did catch him a little accidentally. I don't even think it was a booking.	Shearer thinks it's a poor decision, but doesn't think the result would have changed without the sending off. Lawrenson feels it was harsh on Wigan. They are fighting for position. He puts his arm out to stop Carrick taking the ball away from him and the referee is quite a distance away. You have to be 100% certain it is an intentional thing. In some games you wouldn't even get a foul for that.
			Park Ji-sung goes down for Manchester Utd in the box. Penalty given. Looked like a penalty in real time. Replay shows Park was inside, but foul by Alcaraz took place outside. Definite foul, but shouldn't have been a penalty.	Lawrenson thinks the penalty is harsh against Wigan.

Table A3.1 (continued)

Date	Score	Referee	The Play	Commentators' Discussion
Dec. 26, 2011	Sunderland 1 vs. Everton 1	Howard Webb	Everton's Leon Osman goes down in the Sunderland box and the penalty is given. Looked like a penalty in real time. Replay showed conclusively that there was no contact from Lee Cattermole as I first thought; rather, Leon Osman miskicked and fell over. Wrong decision.	Lineker says it was a rubbish decision. For Lawrenson, this is just a no, no, no, no, no, isn't it? It was just the wrong decision, unfortunately. There's no contact at all for Shearer, and for Howard Webb to give that is a big mistake.
Jan. 1, 2012	Sunderland 1 vs. Manchester City 0	Kevin Friend	Sunderland scores to win game 1–0. Replay shows clear offside. Goal should have been disallowed.	Hartson shows that there is an offside, and points out that the assistant referee is not up with play.
Jan. 2, 2012	Fulham 2 vs. Arsenal 1	Lee Probert	Gervinho goes down in the box and appeals for a penalty to Arsenal. Looked like a penalty in real time. Nothing given. After watching the replay it's a clear penalty, and I can't believe why the referee has not given it.	Shearer thinks this was a penalty.

Table A3.1 (continued)

Date	Score	Referee	The Play	Commentators' Discussion
Jan. 2, 2012	Wolverhampton Wanderers 1 vs. Chelsea 2	Peter Walton	Wolves appeal for a red card as Frank Lampard goes in with a crunching challenge on Hamill. It looked bad in real time, but it's only a yellow card. After watching the replay, he's late, his studs are up, and it's high; he had to go. Poor decision.	Shearer thinks that's a red card.
Jan. 4, 2012	Newcastle 3 vs. Manchester Utd 0	Howard Webb	Ferdinand brings down Ba of Newcastle in the penalty area. It looked like a clear penalty in real time. Replay shows Ferdinand dragged Ba's foot away from him and got nothing on ball. Bad decision.	Hansen thinks that was a penalty kick.
Jan. 4, 2012	Wigan 1 vs. Sunderland 4	Mike Dean	Sunderland's Bendtner goes down and gets a free kick. It looked soft during real time. Replay confirms Bendtner just fell over. Sunderland scores direct from free kick.	Lawrenson feels Martinez has a right to be aggrieved because it's not a free kick. He doesn't see where you give a free kick for that; it's just players battling for the ball.

Table A3.1 (continued)

Date	Score	Referee	The Play	Commentators' Discussion
Jan. 14, 2012	Tottenham 1 vs. Wolves 1	Mike Jones	Corner given to Wolves. Kyle Walker of Spurs complains that he didn't get last touch. Wolves score from the corner. It was difficult to tell in real time, but replay confirms the ball came off the Wolves player last. It shouldn't have been a corner.	Shearer agrees that Edwards touched the ball last and it shouldn't have been a corner.
			Gareth Bale crosses for Adebayor who puts it in the back of the Wolves net. Goal is chopped off for offside, which at first glance seems like the right decision. However, replay shows that Foley at the bottom of the screen may be playing Adebayor onside.	Shearer shows how Kevin Foley is quite clearly keeping Adebayor onside by a yard or more.
Jan. 14, 2012	Chelsea 1 vs. Sunderland 0	Phil Dowd	Chelsea's Fernando Torres is brought down in the Sunderland box. It looks like a penalty in real time, and replay confirms he was brought down by a combination of two tackles.	Hansen thinks that was a certain penalty kick.

Table A3.1 (continued)

Date	Score	Referee	The Play	Commentators' Discussion
			Torres goes down in the box under the challenge of two defenders. Although I think he plays for it, he is brought down and it should be a penalty. He is booked for diving. Poor decision.	Lineker thinks there was contact, and again Phil Dowd got this totally wrong. For Shearer, that angers him because he is basically calling Torres a cheat. If he is going to call him that then he has to be sure, and he can't be sure because there's contact there.
Jan. 21, 2012	Fulham 5 vs. Newcastle 2	Lee Mason	Damien Duff is brought down in the Newcastle penalty area. Referee deliberates and then points to the spot. Looked a bit soft in real time and slightly outside the box. I think it was outside the box. Poor decision.	For Hansen, the referee makes a mistake, because the incident is outside the box. It is a foul, but it's a yard out.

Table A3.1 (continued)

Date	Score	Referee	The Play	Commentators' Discussion
Jan. 21, 2012	Everton 1 vs. Blackburn 1	Mark Halsey	Tim Cahill scores for Everton amidst appeals from the Blackburn defenders for hand-ball. In real time looks like a goal and hard to see what they're on about after multiple angles. However, one final angle shows Fellaini deliberately flick the ball away from the keeper toward Cahill with his hand. Goal shouldn't have stood.	Hansen agrees it's a clear hand-ball.
Jan. 22, 2012	Manchester City 3 vs. Tottenham 2	Howard Webb	Tottenham's Kaboul goes down outside the City box after getting whacked by Lescott. Looked bad in real time. Replay shows he has deliberately elbowed him. Should have been a red card.	Dixon thinks Lescott is lucky to get away with this one. Dixon thinks it should have been a red. Hansen thinks this should be a red card. Might have changed the game.
			Balotelli kicks out at the head of Scott Parker on the ground. He'd already been booked, so that should have been a second yellow, if not a straight red card. Referee didn't have a great view of it, and at first glance it did look like an accidental collision after a challenge. Replay shows it's a deliberate kick.	Dixon says you have to give him the benefit of the doubt. There's reason to believe that because he's off-balance it's hard to be certain. Hansen says it's a stamp. He should have been given a red card.

Table A3.1 (continued)

Date	Score	Referee	The Play	Commentators' Discussion
Jan. 31, 2012	Everton 1 vs. Manchester City 0	Peter Walton	Lescott of Manchester City is clattered into by Everton keeper Tim Howard. It looks a stonewall penalty kick in real time and replay confirms it. Referee gives a foul against Lescott, but he's shoved from behind by defender and then clattered by keeper. Poor decision.	Hansen thinks this is an absolute stonewall penalty. Absolutely no doubt.
			Kolarov crosses the ball and it's blocked by Phil Neville amidst loud appeals from Manchester City players for hand-ball. Looks soft in real time; however, replay shows Neville does move hand out toward ball as he strikes it with his hand. Should have been a penalty.	Hansen thinks because Neville is nine yards away (they've measured it), it's a penalty kick.
Feb. 4, 2012	Stoke 0 vs. Sunderland 1	Martin Atkinson	Robert Huth got sent off for a reckless challenge. Looked bad in real time. From watching the replay, it was late but not dangerous, not two-footed and didn't have studs showing. Should have only been a yellow.	Lawrenson thinks he does try and pull out of the tackle. Martin Atkinson does love a red card—he has the record. Shearer says it's not high, his studs aren't showing: it's not a sending off.

Table A3.1 (continued)

Date	Score	Referee	The Play	Commentators' Discussion
Feb. 5, 2012	Chelsea 3 vs. Manchester Utd 3	Howard Webb	Manchester Utd appeals for a penalty as Wellbeck goes down under a challenge from Cahill. Looked like a penalty in real time. Replay shows it was a foul, but I think it was just outside the box. However, if it was a foul then Cahill could also have been sent off for denying a goal-scoring opportunity.	Dixon thinks this is a foul, and could have been a red card.
Mar. 3, 2012	Liverpool 1 vs. Arsenal 2	Mark Halsey	Liverpool's Suárez is brought down by Arsenal goalkeeper in the box. Penalty given. Looked like a penalty in real time. Replay suggests there was minimal contact.	Shearer thinks he's made the most of it. When he first saw it, Shearer agreed with the referee. He doesn't think referee has a great view of it. Shearer thinks it's merely the brushing of a sock. He's made the most of it. Hansen agrees there's no contact there.
Mar. 3, 2012	Stoke 1 vs. Norwich City 0	Michael Oliver	Stoke get throw-in, which appears to come off their own player Marc Wilson last. Stoke score straight from the throw-in. Replay confirms it should have been a Norwich City throw.	Shearer agrees the linesman got it wrong. It's a poor decision.

Table A3.1 (continued)

Date	Score	Referee	The Play	Commentators' Discussion
Mar. 10, 2012	Bolton 2 vs. Queens Park Rangers 1	Martin Atkinson	QPR defender Clint Hill headers toward goal, and appeals that it crossed the line. Not given. It looked in real time that it went over the line. Replay confirms it was well over the line. Referee had a decent position. Really bad decision.	Lawrenson feels it is inexplicable. There's no argument whatsoever. Lawrenson says look at all the other sports that use it (replay). They laugh at us; it's a billion dollar industry and they laugh at us. Lineker says if he can't see it he can't see it (making reference to the fact that the linesman signaled he had two people in front of him). Hansen says it might have led to QPR winning the game, taking momentum into the next game and so on. Effects are incalculable.
Mar. 11, 2012	Manchester Utd 2 vs. West Bromwich Albion 0	Lee Probert	Odemwingie goes down in the box and West Bromwich Albion appeals for a penalty. Looked like a foul at the time. Replay shows Evra's foot was out and he caught him. Poor decision.	Hansen thinks that this is a penalty.

Table A3.1 (continued)

Date	Score	Referee	The Play	Commentators' Discussion
Mar. 17, 2012	Wigan 1 vs. West Bromwich Albion 1	Michael Oliver	Wigan's James McArthur chases and brings down West Bromwich Albion's Mulumbu. He gets the foul but then gets up, chases the Wigan player, and pushes him. He is only booked, but that should have been red. His arms were raised.	Hansen thinks he was very lucky to stay on the pitch. If you raise your arms, the letter of the law says you're off the pitch. That might have changed the game.
Mar. 21, 2012	Manchester City 2 vs. Chelsea 1	Mike Dean	Penalty given to Manchester City. Zabaleta strikes the ball and it hits the hand of Chelsea's Michael Essien. Looks like a penalty in real time. Replay shows Essien put his hands up in the air, but the ball was only two yards from him. I'm not sure what he could have done to avoid it. Tough call, but I think he made the wrong decision.	Hansen says you can argue that if you put your hands up in the air you're in trouble, but I don't think that's a penalty kick at all. Dixon can see why it's given, but I think he's unlucky. Sort of penalty where if you're defending you don't think it is a penalty, and when you're attacking you think it is a penalty.
Mar. 21, 2012	Everton 0 vs. Arsenal 1	Lee Mason	Everton's Drenthe scores, but goal is chopped off for offside. Looks harsh in real time. Replay confirms he is a yard onside. Poor decision.	Dixon says it's a bad decision by the linesman.

Table A3.1 (continued)

Date	Score	Referee	The Play	Commentators' Discussion
Mar. 24, 2012	Bolton 2 vs. Blackburn 1	Andre Marriner	Blackburn's Hoilett runs into the box and goes to ground. No penalty given. Looked like a good shout in real time. Replay confirmed there was definite contact, and I think it should have been a penalty.	Dixon thinks it was a blatant penalty and he doesn't know how the referee failed to give it. Andre Marriner had a good position. Shearer agrees it was a blatant penalty. Referee has made a mistake there.
Mar. 24, 2012	Stoke 1 vs. Manchester City 1	Howard Webb	Gareth Barry of Manchester City went in on a thumping challenge against the Stoke striker. Looked bad in real time and could have been a penalty for Stoke—maybe even a red card.	Dixon feels it was a penalty. It's a foul and it's inside the box. He presumes the referee thought he got the ball; but he didn't.
Mar. 31, 2012	Manchester City 3 vs. Sunderland 3	Phil Dowd	Mario Balotelli of Manchester City gets pulled down in the box. Looks like a penalty in real time, but nothing given. Replay shows Turner was holding Balotelli, and I've seen them given.	Hansen thinks this was a penalty kick.

Table A3.1 (continued)

Date	Score	Referee	The Play	Commentators' Discussion
Apr. 1, 2012	Newcastle Utd 2 vs. Liverpool 0	Martin Atkinson	Liverpool appeal that ball crossed the Newcastle goal line from a corner. Looked close in real time. Replay shows it didn't cross the line, but, rather, Simpson handles it. Clear penalty and also would have been a red card for denying goal.	Jason Roberts says the referee can't see this incident, but it's a clear hand-ball and he would have been sent off. Lawrenson says that would have been key, as you could be 1–0 up playing against ten men.
			Cissé scores for Newcastle, amidst appeals from Liverpool players that he was offside. Replay shows he was offside at different passages of play. Poor decision.	Lawrenson says there was a strong suspicion of offside.
Apr. 7, 2012	Chelsea 2 vs Wigan 1	Mike Jones	Ivanović scores for Chelsea, but looks marginally offside when doing so. Wigan players surround the referee and after deliberating with his linesman he gives the goal. Replay shows Ivanović was about two yards offside. Really poor decision by the officials.	Hansen says they've been undone by two bad decisions that were really poor. He's miles offside.

Table A3.1 (continued)

Date	Score	Referee	The Play	Commentators' Discussion
			Mata scores for Chelsea. Looked like a good goal in real time. Replay shows that Mata was offside when Torres hit toward back post. Poor decision.	Hansen shows a camera angle that confirms that Mata is offside. Surely the linesman has got to see that? It's cost Wigan dearly.
Apr. 7, 2012	Liverpool 1 vs. Aston Villa 1	Michael Oliver	Suárez goes down in the box under the challenge of Villa's Alan Hutton. Looked like a penalty in real time and replay confirms there was contact from Hutton. Felt Suárez was hard-done by on this occasion.	Lawrenson feels Suárez gets clipped there, but because of his reputation he doesn't get the decision. He thinks it is a penalty, regardless of what you think about Suárez.
			Kuyt of Liverpool chests ball in the box and it strikes the arm of the Aston Villa defender who had his back to play. His arms are out by his side and I think it's a clear penalty. Nothing given.	Lawrenson thinks this was a penalty, albeit he sympathizes with the fullback as he thinks he's putting his arms out to feel for Kuyt.
			Carruthers of Aston Villa is brought down on the very edge of the Liverpool box by Daniel Agger. Referee books him for simulation. Replay shows he was caught by Agger and it should have been a penalty.	Lawrenson thinks this is harsh on Carruthers, because he thinks this is a penalty.

Table A3.1 (continued)

Date	Score	Referee	The Play	Commentators' Discussion
Apr. 8, 2012	Manchester Utd 2 vs. Queens Park Rangers 0	Lee Mason	QPR defender Derry sent off for denying a goal-scoring opportunity to Ashley Young by bringing him down for a penalty. Looked like good decision in real time, but replay shows Young was a yard offside when ball was played through to him. Should have been a free kick to QPR. Poor decision— especially as linesman is looking right above the line. On top of this Young has gone down far too easily—as he consistently does.	Dixon says this decision is unfathomable. He's looking straight along the line, and he can't believe how he didn't see it. Decisions like that change games. Dixon thinks the referee has been fooled by Young. Savage also thought it was minimal contact. He's dived, and this isn't the first time Young has done it. Savage himself admits that he's cheated. He says the referee could not wait to blow his whistle and send him off. Savage says there are degrees of cheating. Colin Murray says it's a big club getting a decision against a little club.

Table A3.1 (continued)

Date	Score	Referee	The Play	Commentators' Discussion
Apr. 8, 2012	Arsenal 1 vs. Manchester City 0	Martin Atkinson	Shocking challenge by Balotelli on Arsenal's Alex Song. Referee doesn't see it. It's late, above the knee and I think deliberate. Should have been a red card. Can't believe the referee missed it.	Savage says that's a leg-breaking challenge, and he should have been sent off.
Apr. 9, 2012	Tottenham 1 vs. Norwich City 2	Michael Oliver	Tottenham's Ledley King wrestles Grant Holt to the ground in the box. Looked a stonewall penalty in real time and replay confirmed this. Poor decision not to give it.	Lawrenson says this is a penalty all day long. They were right to feel aggrieved about it.
			Wilbraham of Norwich City goes down in the box under a clumsy challenge from Adebayor. Looked like a penalty in real time. Replay confirms it.	Lawrenson says this is a foul anywhere else on the pitch, so why hasn't the referee given it?
Apr. 9, 2012	Newcastle 2 vs. Bolton 0	Mike Jones	Cissé scores for Newcastle. Looked like a good goal in real time. Replay shows there was a hint of offside about the goal as Cissé appeared to be in front of the last man when Ameobi crossed it.	Hansen thinks he looked offside.

Table A3.1 (continued)

Date	Score	Referee	The Play	Commentators' Discussion
Apr. 9, 2012	Fulham 1 vs. Chelsea 1	Mark Clattenburg	Chelsea's Salomon Kalou goes down in the penalty box. I thought it looked soft in real time and like he played for it. Replay shows it was touch and go.	Lawrenson says when you have to magnify a tackle in the box to see a foul you know you're in trouble. Hansen agrees it is soft. He says they have watched it fifteen times and if there is any contact at all he's not sure where it is.
Apr. 9, 2012	Aston Villa 1 vs. Stoke 1	Mike Dean	Jermaine Pennant goes down under a challenge from Gardner of Aston Villa. Looks really soft in real time and replay confirms this. Stoke score from the resulting free kick, where Shawcross seems to be climbing on James Collins also.	Hansen doesn't think that was a legitimate foul and also feels Shawcross fouled James Collins.

Table A3.1 (continued)

Date	Score	Referee	The Play	Commentators' Discussion
Apr. 11, 2012	Wigan 1 vs. Manchester Utd 0	Phil Dowd	Victor Moses scores for Wigan. There is then mass confusion as music plays, and players celebrate; goal is chopped off for foul on goalkeeper. Decision seems to have come from the linesman, rather than the referee. Can't see much of it in real time. Replay shows that Caldwell is standing in front of the keeper, but de Gea does almost nothing to get out of the way. Looks like a really poor decision.	Hansen doesn't think there was anything wrong here. Caldwell really isn't doing anything.
			Manchester Utd's Phil Jones and Wigan player tussle on the bye line. Corner given to Wigan, but Jones argues vehemently that it came off Wigan player last. Wigan score from corner. Multiple camera angles are played, and eventually it does seem that ball came off Wigan player last. Corner shouldn't have been given, but was tough call.	Hansen thinks this is a bye kick to Manchester Utd. Poor decision.

Table A3.1 (continued)

Date	Score	Referee	The Play	Commentators' Discussion
			Johnny Evans hacks down Maloney of Wigan on halfway line. He's already been booked and should have been sent off.	Hansen thinks he should be red carded for this challenge.
			Phil Jones crosses into the Wigan box. Ball is blocked by Figueroa, amidst strong appeals from Manchester Utd players for offside. Replay confirms it did strike his hand and it should have been a definite penalty.	Hansen thinks this is stonewall penalty.
Apr. 11, 2012	Queens Park Rangers 3 vs. Swansea 0	Lee Probert	Jamie Mackie of QPR is threaded in by teammate to score. Looked like a good goal in real time. Replay shows he was marginally offside when ball was played. Goal shouldn't have stood.	Martin Keown thinks he's definitely offside when ball is played.
Apr. 14, 2012	Norwich City 1 vs. Manchester City 6	Chris Foy	Manchester City's Tevez is booked for simulation in the box. Looked like a penalty in real time I thought, but replay shows the Norwich City defender made contact with Tevez's foot. I think that's a clear penalty. Poor decision.	Shearer thinks in no way this was a dive. Referee has a great position, but just got it completely wrong. Defender stands on his foot.

Table A3.1 (continued)

Date	Score	Referee	The Play	Commentators' Discussion
Apr. 15, 2012	Manchester Utd 4 vs. Aston Villa 0	Mark Halsey	Penalty to Manchester Utd as Ashley Young goes down again in the box. Looked like a dive in real time. Replay shows Young has quite deliberately left his leg out to touch the player's trailing leg. That's simulation in my estimation.	Dixon says Ashley Young went down easily. Dixon said the defender is trying to pull his foot away. Very, very soft penalty. Olsson thinks there is contact, but it's Young stepping on his foot.
Apr. 21, 2012	Aston Villa 0 vs. Sunderland 0	Anthony Taylor	Bendtner scores for Sunderland, but it's been chopped off for offside. Goal looked good in real time. Replay confirms Bendtner was marginally onside.	Shearer believes he wasn't offside at all. Whatever happened to giving the forward the advantage? Uses technology line to demonstrate this.
Apr. 28, 2012	Everton 4 vs. Fulham 0	Phil Dowd	Jelavić scores Everton's third goal. He looked marginally offside in real time. Replay confirms he was offside. Poor decision. Goal shouldn't have stood.	Lawrenson thought he was offside.
Apr. 28, 2012	West Bromwich Albion 0 vs. Aston Villa 0	Mark Clattenburg	Ball appears to strike the hand of Chris Brunt in the penalty area. Replay shows he didn't know much about it, but his arm was out and I've seen them given.	Lawrenson thinks this should have been given. Referee had a good view, so he's not sure why he didn't give it.

Table A3.1 (continued)

Date	Score	Referee	The Play	Commentators' Discussion
			West Bromwich Albion appeal for hand-ball from corner kick. Liam Ridgewell headers toward goal and Villa defender Hutton sticks his arm out and flicks it over the bar. It should have been a penalty, and possibly a red card as Hutton stopped ball from going in the net I think.	Lawrenson thinks that's a penalty all day long. He'd already been booked as well so he would have been sent off. Linesman had a great view of it.
			Villa player slides ball toward goal, but it's blocked by sliding West Bromwich Albion defender. Villa appeal for hand-ball, but it's hard to tell in real time what happened. Replay shows the West Bromwich Albion defender Olsson slid in and handled it on the ground.	Lawrenson thinks Olsson scoops this and the referee had a great view there also. Clear-cut penalty.
May 13, 2012	Stoke 2 vs. Bolton 2	Chris Foy	Stoke's Walters jumps into the Bolton goalkeeper as the ball hangs in the air. He bundles it in the net, and the goal stands. Looked a foul in real time. Replay shows it was a foul, as the keeper already had the ball in his hands. Poor decision.	Shearer thinks that this was without doubt a foul. Keeper has it in both hands. You might have seen that forty years ago, but he's just barged it out of his hands. That's a foul.

Table A3.2
Wrong refereeing decisions: Season 2012–2013

Date	Score	Referee	The Play	Commentators' Discussion
Aug. 19, 2012	Manchester City 3 vs. Southampton 2	Howard Webb	Tevez scores amidst appeals from Southampton that he was offside. Looked tight in real time. Replay showed it was tight. I think he was slightly offside.	Mark Lawrenson thinks there was a sniff of offside about this goal.
Aug. 25, 2012	Southampton 0 vs. Wigan 2	Anthony Taylor	Appeals for foul from Southampton as Koné wrestles ball from Fonte and goes on to score. I thought it was a foul at the time. Replay shows he did stand on Fonte's heel. Should have been a foul and no goal.	Lawrenson thinks it was a foul.
Sept. 1, 2012	Manchester City 3 vs. Queens Park Rangers 1	Chris Foy	Bobby Zamora scores for Queens Park Rangers on the rebound after keeper saves Cissé's penalty. Replay shows he was in the box already when ball was struck. The penalty should have been retaken.	Alan Hansen thinks Zamora is encroaching and it should have been retaken.
Sept. 1, 2012	Tottenham 1 vs. Norwich City 1	Mark Halsey	Assou-Ekotto of Spurs holds Norwich City's Jordan in the penalty area. Looked like a penalty. Replay shows it was a clear-cut penalty. Really poor decision.	Alan Shearer thinks it's a stonewall penalty. Redknapp thought this was a penalty.

Table A3.2 (continued)

Date	Score	Referee	The Play	Commentators' Discussion
			Tottenham's Huddlestone is sent off for dangerous play as he brings down Howson on halfway line. Was a bad, two-footed challenge. I'm not sure it was a red as both players slid in with two feet, but have seen them given.	Harry Redknapp thought this was a harsh sending off. His feet weren't off the ground, and he wasn't going in to hurt anybody.
Sept. 15, 2012	Manchester Utd 4 vs. Wigan 0	Michael Oliver	Danny Welbeck of Manchester Utd goes down under challenge from Wigan keeper. Looked soft in real time. Replay shows it was a soft penalty, as I don't think there was a lot of contact. Keeper even pulled out of the challenge and slid in with his knees instead.	Shearer thinks the keeper is stupid here, but it's a dive from Welbeck—he should have been booked for this.
Sept. 15, 2012	Stoke City 1 vs. Manchester City 1	Mark Clattenburg	Manchester City players appeal Peter Crouch handled ball in build-up to his goal. Hard to tell in real time. Replay confirmed it was a hand-ball. Poor decision. Goal should have been chopped off.	Hansen thinks it's a double hand-ball.
Sept. 15, 2012	Queens Park Rangers 0 vs. Chelsea 0	Andre Marriner	John Terry goes down as he is held by Ryan Nelsen. Looked like a penalty after viewing the replay. Nelsen had his hands all over him. Poor decision.	Shearer thinks this is a rugby tackle.

Table A3.2 (continued)

Date	Score	Referee	The Play	Commentators' Discussion
			Chelsea's Eden Hazard appeals for penalty after he is brought down by Wright-Phillips. Looked like a penalty in real time. Replay confirms it was a penalty. Poor decision, as Wright-Phillips got nothing on the ball.	Shearer thinks this is a penalty. Referee has a great position, but doesn't give it for some reason.
Sept. 16, 12	Reading 1 vs. Tottenham 3	Howard Webb	Reading claim hand-ball by Kyle Walker. Did look like a hand-ball. Referee felt Walker was fouled as he went for ball, so gave free kick to Spurs instead. I thought it should have been a penalty.	Robbie Savage thinks this is a massive turning point in the game. He shows the referee's view of the incident, and says it's a blatant hand-ball. Every Reading player appeals. Savage thinks it's never a push. Dion Dublin thought it was definitely a penalty.
Sept. 22, 2012	Swansea 0 vs. Everton 3	Anthony Taylor	Anichebe scores for Everton. Replay shows that Fellaini elbowed the ball into the build-up to the goal. The goal should have been disallowed and a free kick given to Swansea.	Redknapp thinks this is a hand-ball, and they've been fortunate there.

Table A3.2 (continued)

Date	Score	Referee	The Play	Commentators' Discussion
Sept. 22, 2012	Chelsea 1 vs. Stoke City 0	Michael Oliver	David Luiz goes in two-footed on Stoke's Jon Walters. Bad challenge and I think it should have been a red card. He was only booked.	Lawrenson thinks it's a straight red card. Any player should get a red card for this. Redknapp says that's a horrendous tackle. It's a leg-breaker.
Sept. 29, 2012	Manchester Utd 2 vs. Tottenham 3	Chris Foy	Nani goes down in the box amidst appeals for a penalty. Looked like he was being held in real time. Replay suggests it was a definite penalty—although Nani went down a little theatrically. Poor decision.	Mick McCarthy doesn't think the referee had a great view of it, and it's up to the assistant therefore to give it. Clear penalty. Hansen thinks this is a stonewall penalty.
Sept. 29, 2012	Norwich City 2 vs. Liverpool 5	Mike Jones	Suárez goes down in box and appeals for penalty. Looked like a foul in real time. Replay shows Suárez was bundled to the ground and it's a clear penalty. Commentator notes that perhaps Suárez's reputation has preceded him in this instance.	McCarthy says if that's not a penalty then he doesn't know what is.

Table A3.2 (continued)

Date	Score	Referee	The Play	Commentators' Discussion
Sept. 29, 2012	Reading 2 vs. Newcastle Utd 2	Andre Marriner	Ba scores for Newcastle amidst appeals that he hand-balled it into the goal. Was very difficult to see in real time and from the referee's perspective. Replay shows it struck his head and then off his hand. The goal shouldn't have counted.	Hansen thinks it's difficult for referee to see it, but it's a definite hand-ball.
Oct. 6, 2012	Wigan 2 vs. Everton 2	Kevin Friend	Koné scores for Wigan amidst Everton appeals he was offside. I thought he looked off in real time, and replay shows he was offside. Really poor decision by the assistant referee.	Martin Keown thinks he was offside, so the goal shouldn't have counted.
Oct. 7, 2012	Liverpool 0 vs. Stoke City 0	Lee Mason	Robert Huth stomps on chest of Suárez. Nothing given by the referee. Replay shows he does stand on the chest of Suárez. Hard to tell whether it was deliberate. If it was he should have been sent off.	Jason Roberts says it's a definite stamp. Could have been a red card.

Table A3.2 (continued)

Date	Score	Referee	The Play	Commentators' Discussion
Oct. 20, 2012	Swansea 2 vs. Wigan 1	Mike Jones	Koné scores for Wigan to equalize, but it's chopped off for offside. It looked tight at the time. Replay shows he was level. Poor decision by the assistant referee.	Redknapp feels it is incredibly tight. Shearer thinks it's also incredibly tight, and Wigan is hard-done by. It's a big call by the assistant referee. Swansea manager Michael Laudrup just saw the replay and he thought it was tight and perhaps they were lucky.
Oct. 21, 2012	Queens Park Rangers 1 vs. Everton 1	Jon Moss	Jelavić appeals for a penalty for Everton. Looked like a penalty in real time. Replay shows he was clearly fouled by Mbia. Stonewall penalty.	Shearer thinks this was a penalty. McCarthy also thinks this was a penalty.
			Pienaar is sent off for second yellow card. The first yellow was deserved, but I think the second yellow was very harsh. If anything Bosingwa makes contact with Pienaar's foot when he swings foot back to kick the ball.	Shearer doesn't think the second challenge was a yellow card. It looks like a little clip, but when you slow it down Bosingwa is the one who kicks Pienaar's leg. Colin Murray mentions that everyone thought it was sending off when they watched at the time; referees don't have that luxury of watching replays.

Table A3.2 (continued)

Date	Score	Referee	The Play	Commentators' Discussion
			Queens Park Rangers appeal for a penalty, as Coleman bundled Hoilett down to the ground. Looked like a penalty I thought at the time. Replay shows it's a stonewall penalty. Really poor decision.	Shearer thinks that's a penalty, and the referee has got all three decisions wrong. McCarthy says if he's the referee that's a penalty.
Oct. 27, 2012	Arsenal 1 vs. Queens Park Rangers 0	Anthony Taylor	Arteta scores for Arsenal amidst appeals he was offside. Replay shows Arteta was standing on the line, and I think he is offside.	Hansen thinks Arteta is offside and Queens Park Rangers are very unfortunate to lose that goal.
Oct. 27, 2012	Aston Villa 1 vs. Norwich City 1	Phil Dowd	Norwich City appeals that Villa captain Vlaar pushes forward in the box. Looked like as clear penalty in real time and replay confirms it was a poor decision.	Hansen thinks this was a stonewall penalty. Referee is right there, so doesn't know why he doesn't give it.
			Aston Villa's Herd fouls Norwich City player. Bad challenge, should have been a second yellow card and a sending off. Poor decision not to book him.	Hansen thinks he should have been sent off.
Oct. 27, 2012	Stoke City 0 vs. Sunderland 0	Mark Halsey	Sunderland's Fletcher strikes at goal and claims for hand-ball by Robert Huth. Looked like a penalty after watching the replay.	Lawrenson feels that this was a penalty. Gary Lineker says the shot was going in until it was blocked.

Table A3.2 (continued)

Date	Score	Referee	The Play	Commentators' Discussion
Oct. 28, 2012	Chelsea 2 vs. Manchester Utd 3	Mark Clattenburg	Chelsea's Fernando Torres is sent off for second yellow, as referee deemed him to have dived after challenge by Evans. His first booking was the right decision. Replay shows Evans caught him. Poor decision.	Hansen thinks the first one is a yellow, but not the second one. Graeme Le Saux says there is a touch on his leg, but it's a delayed fall.
			Hernandez scores for Manchester Utd. Replay shows he was in an offside position when the ball was played. Goal shouldn't have stood.	Hansen shows that he is definitely offside. Le Saux says assistant referee's view is being blocked.
Oct. 28, 2012	Everton 2 vs. Liverpool 2	Andre Marriner	Liverpool appeals that an Everton throw-in should have been their throw-in, as last touch came off Raul Meireles. Everton score from the throw-in. Was hard to see in real-time who touched it last. Replay shows it came off Meireles. Poor decision.	Le Saux says it is only in replay that you see it, but it should have been a Liverpool throw-in.
			Suárez scores for Liverpool, but it's chopped off for offside. Looked tight in real time. Replay shows he was nowhere near offside. Shocking decision.	Hansen shows there's no debating that Suárez is onside. The assistant referee doesn't flag until very, very late as well—strange decision.

Table A3.2 (continued)

Date	Score	Referee	The Play	Commentators' Discussion
Nov. 3, 2012	Norwich City 1 vs. Stoke City 0	Andre Marriner	Stoke is furious as Adam is fouled, but referee doesn't give free kick. Norwich City goes straight up the other end, wins a free kick, which they subsequently score from. It looked like a free kick, and replay confirms it was a poor decision. (Adam had been booked for diving earlier, and I don't know if this played on the referee's mind.)	Hansen thinks this is a free kick.
Nov. 3, 2012	West Ham Utd 0 vs. Manchester City 0	Howard Webb	Kevin Nolan scores for West Ham but it's chopped off for offside. Looked like good goal in real time. Replay shows he was onside. Poor decision by the assistant referee.	Shearer thought this was the wrong decision, and a good goal. Hansen shows how Džeko is playing him onside.
Nov. 10, 2012	Southampton 1 vs. Swansea 1	Andre Marriner	Lallana goes down in the box and no penalty given. This is a poor decision, as there is clearly contact from Garry Monk. Should have been a penalty.	Keown thinks this is a definite penalty. It's almost criminal; how the referee has missed that, he can't believe it.
Nov. 11, 2012	Chelsea 1 vs. Liverpool 1	Howard Webb	Chelsea appeal that Suárez pushes Ramires before heading in for Liverpool. He did push him, although he was being held. Goal shouldn't have counted.	Hansen says it's conclusive. No goal. Roberts thinks it's a blatant push.

Table A3.2 (continued)

Date	Score	Referee	The Play	Commentators' Discussion
Nov. 11, 2012	Newcastle Utd 0 vs. West Ham Utd 1	Mike Dean	Newcastle appeal that Ba's header crosses the line before keeper saves it. Looked like it did in real time. Replay shows the ball did cross the line.	Despite the fact that the pundits didn't discuss it, I'm including this decision in my analysis, as it was clearly well over after looking at replay.
Nov. 17, 2012	Manchester City 5 vs. Aston Villa 0	Jon Moss	Strange incident. Corner comes in from Manchester City, is blocked by Villa's Vlaar, and goes behind for another corner. Suddenly the assistant referee raises his flag and referee gives a penalty. No idea what it's been given for in real time. Replay suggests that assistant referee gave penalty, because he thought he saw hand-ball by Villa's Weimann. It doesn't look like he makes contact with his hand, however.	Lineker thinks this was an awful decision that went Manchester City's way.
Nov. 17, 2012	Liverpool 3 vs. Wigan 0	Kevin Friend	Liverpool's Suárez stamps on the Wigan defender miles away from the ball. Poor challenge, and he should have been sent off.	Hansen thinks this could have been a red card.

Table A3.2 (continued)

Date	Score	Referee	The Play	Commentators' Discussion
Nov. 24, 2012	Sunderland 2 vs. West Bromwich Albion 4	Mike Dean	Penalty to West Bromwich Albion as Johnson brings down Ridgewell. Looked very soft in real time. Replay showed Johnson didn't get anything on the ball, but I think there was no contact whatsoever with Ridgewell; certainly nowhere near enough to bring him down.	Lineker doesn't think there was much contact. He says if this was Suárez they'd be up in arms about a foreign player diving, etc. It's difficult for referees, they don't have all the angles we have. Michael Owen thought at first it was a penalty, especially from the referee's angle. But a replay showed that he dived for that. He waited for contact. Dublin says there's no contact.
Nov. 27, 2012	Wigan 3 vs. Reading 2	Howard Webb	Reading appeal for penalty as Tabb goes down under challenge from Figueroa. Looked like a stonewall penalty in real time. Replay confirms it was a poor decision, especially given he had a good view of the incident.	Lineker thinks this was a penalty. Dublin thinks referee got this completely wrong. He doesn't have all of the replays that we have. Owen thinks there was definite contact.
			Wigan's Di Santo scores, but it's chopped off for offside. Looked tight in real-time. Replay shows it was very unfortunate for Wigan. He was marginally onside. Poor decision.	Lineker says this was a decent goal that was wrongly chopped off.

Table A3.2 (continued)

Date	Score	Referee	The Play	Commentators' Discussion
Nov. 24, 2012	Everton 1 vs. Norwich City 1	Mike Jones	Norwich City gets a free kick. Everton appeals that it wasn't a foul. Norwich City scores from free kick to make it 1–1. It looked soft in real time, and replay suggests there was a slight push, but I think Morrison slips.	Dublin doesn't think it was a free kick. Player goes down far too easy. Lineker also doesn't think it should have been a free kick.
Nov. 25, 2012	Chelsea 0 vs. Manchester City 0	Chris Foy	David Luiz tugs, and holds Kompany for the entire corner kick. Should have been a penalty. Replay shows it was laughable that the referee didn't give a penalty to Manchester City—although holding is becoming rampant in the game.	Peter Schmeichel says if that is not a penalty then he doesn't know what is. He doesn't think it should be part of the game. Give penalties and it will stop. Shearer says it's a penalty, and referees need to start doing something about this to eradicate it from the game.
Nov. 25, 2012	Swansea 0 vs. Liverpool 0	Jon Moss	José Enrique scores for Liverpool, but it's chopped off for offside. Looked like the right decision in real time. Replay shows he is just about level. You could argue the assistant referee should favor the attacker.	Shearer thought this was harshly disallowed for offside. He thinks he was onside. Shearer: "Sometimes you get them, sometimes you don't. They even themselves out." Colin Murray: "Couldn't agree more. No need for any video replays."

Table A3.2 (continued)

Date	Score	Referee	The Play	Commentators' Discussion
Nov. 28, 2012	Everton 1 vs. Arsenal 1	Michael Oliver	Steven Pienaar appeals for penalty, as Arsenal's Arteta brings him down in the box. I thought it was touch-and-go after watching the replay. Have seen them given though.	Hansen thought this was a definite penalty. Shearer felt from the referee's angle he thinks Arteta touched it. Should have been a penalty.
Dec. 1, 2012	Reading 3 vs. Manchester Utd 4	Mark Halsey	Manchester Utd appeals that van Persie's shot crosses the line. It looked like it did in real time. Replay shows it was miles over. Really poor decision by the assistant referee.	Pundits didn't discuss, but the ball was at least a foot over the line after watching the replay.
Dec. 8, 2012	Swansea 3 vs. Norwich City 4	Howard Webb	Swansea scores, but the referee chops it off for a foul on the Norwich City keeper. Looked harsh on Swansea.	Hansen thinks they are desperately unlucky. He thinks it's a goalkeeping error.
Dec. 8, 2012	Sunderland 1 vs. Chelsea 3	Mark Halsey	Hazard is pulled back in the Sunderland box, and appeals for penalty. Replay shows he was definitely pulled back. It was a stonewall penalty. Poor decision.	Shearer says that should have been a penalty.
Dec. 8, 2012	Arsenal 2 vs. West Bromwich Albion 0	Mike Jones	Penalty to Arsenal as Cazorla goes down in the box. Looked like the right decision in real time. Replay shows there was no contact. He's dived and that's a poor decision.	Lineker thinks it's not a penalty. Hansen thinks this is an embarrassment. There's no contact whatsoever.

Table A3.2 (continued)

Date	Score	Referee	The Play	Commentators' Discussion
			Arsenal gets a second penalty, but West Bromwich Albion appeals that there was a foul by Oxlade-Chamberlain in the build-up. I thought it was a foul when I first saw it. Replay shows the penalty was the right decision, but he did foul Popov in the build-up. Poor decision not to give the free kick to West Bromwich Albion.	Hansen thinks this is a clear foul. Poor decision. Steve Clarke is right in saying that the referee has had a poor afternoon.
Dec. 8, 2012	Aston Villa 0 vs. Stoke City 0	Roger East	Ryan Shotton is sent off for two yellow cards. The second yellow was farcical. Poor decision.	Hansen thinks this was a poor decision. Shearer thinks this was a staggering decision to book him.
Dec. 9, 2012	Manchester City 2 vs. Manchester Utd 3	Martin Atkinson	Ashley Young scores for Manchester Utd after shot comes off post. Assistant referee chops it off for offside. Replay shows he was in line. Poor decision.	Hansen says he was clearly onside and the game should have been over at 3–0.
Dec. 15, 2012	Newcastle Utd 1 vs. Manchester City 3	Andre Marriner	Carlos Tevez is played through and slots it into the net. Goal is chopped off for offside. Replay confirms he was definitely onside. Goal should have counted.	Shearer agrees he isn't offside.

Table A3.2 (continued)

Date	Score	Referee	The Play	Commentators' Discussion
Dec. 15, 2012	Stoke 1 vs. Everton 1	Mark Halsey	Stoke's Richard Shawcross goes down in his own box. Replay shows he is head-butted by Marouane Fellaini. It should have been a sending off for the Everton player. Nothing was given.	Lawrenson thinks it was an act of thuggery. Fellaini should have the book thrown at him. He looks to see where the referee is before he butts Shawcross. He thinks referee is distracted. Shearer thinks there were three incidents in thirty minutes that Fellaini should have been sent off for.
Dec. 22, 2012	Manchester City 1 vs. Reading 0	Mike Dean	The ball strikes City defender's hand as it goes into the box, but penalty not given; play goes on and then Reading forward is brought down in the box by Ricketts. Defender appears to hold his running line, but he definitely brought the player down. Bad decision.	Savage thinks the hand-ball is a penalty. Hansen also thinks it's a penalty.
Dec. 22, 2012	West Ham Utd 1 vs. Everton 2	Anthony Taylor	Carlton Cole sent off for high-foot on Leighton Baines. It's an awful high-foot, but I don't think there's intent to hurt player as he's looking at ball. Should have been a yellow card.	Big thing for Hansen is that Cole's eyes never deviate from the ball.

Table A3.2 (continued)

Date	Score	Referee	The Play	Commentators' Discussion
			Darron Gibson is sent off for a high foot on Noble. It wasn't a red card, but referee had no choice after the previous decision.	The referee has to even it up, but Hansen thinks both those red cards should be rescinded. Savage thinks if both of those red cards aren't rescinded, then the whole disciplinary panel should be sacked. Simple as that.
Dec. 22, 2012	West Bromwich Albion 2 vs. Norwich City 1	Martin Atkinson	Goal for West Bromwich Albion from corner. Replay showed it wasn't a corner to begin with. After this, Odemwingie appeared to be backing into the goalkeeper and I think it might have been a foul.	Savage doesn't think it was a corner. He thought it was a free kick also. Hansen notes that it's OK for them because they are able to look at it 20–30 times afterward, whereas the referee only has one chance. Gabby Logan says there are going to be discrepancies. Decision making is different in every game. When *Match of the Day* was on fifty years ago, they didn't have cameras everywhere.

Table A3.2 (continued)

Date	Score	Referee	The Play	Commentators' Discussion
Dec. 22, 2012	Wigan 0 vs. Arsenal 1	Jon Moss	Gomez of Wigan strikes toward goal and it's blocked by Arsenal's Kieran Gibbs. Wigan appeal for hand-ball. Replay shows they have a shout, but not conclusive.	For Hansen, his arms are there. Wigan is hard done by and are desperately unlucky.
Dec. 22, 2012	Southampton 0 vs. Sunderland 1	Howard Webb	Ayuka is brought down by Cuéllar in the box. Looks like it could have been a penalty to Southampton. Replay shows there was contact, but I think he was looking for it. Still probably a penalty.	Savage thinks this was a penalty.
Dec. 26, 2012	Sunderland 1 vs. Manchester City 0	Kevin Friend	Zabaleta is fouled on the left touchline in the build-up to Adam Johnson's goal. It looked like a foul in real time and replay confirms a foul should have been given. Tevez stopped playing, etc.	Lawrenson feels like it was definitely a foul. He can't understand why referee didn't give the foul.
Dec. 26, 2012	Queens Park Rangers 1 vs. West Bromwich Albion 2	Chris Foy	West Bromwich Albion defender Ridgewell clearly handles the ball. Looks like a penalty in real time and replay confirms his arm was up in the air and the ball struck his hand. Definite penalty.	Shearer thinks there is a strong case for a penalty. You raise your arm and you risk giving away a penalty. Poor decision and referee should have received help from his assistant.

Table A3.2 (continued)

Date	Score	Referee	The Play	Commentators' Discussion
Dec. 26, 2012	Everton 2 vs. Wigan 1	Lee Mason	Shaun Maloney of Wigan goes down in the box under the challenge of Leon Osman. Looks like a penalty in real time. Replay confirms Osman left his foot out and caught Maloney.	Lawrenson thinks it was an obvious penalty. He stuck his leg out. No matter how many times you see it, Osman sticks his leg out. If referee doesn't think it's a penalty, then he needs to book him. Lineker doesn't think Wigan is getting the breaks at the minute.
Dec. 29, 2012	Stoke 3 vs. Southampton 3	Mark Clattenburg	Loud appeals from Kenwyne Jones that the Southampton defender, Fonte, handled the ball. Replay confirms his arm was high in an unnatural position and he punched it clear. Should have been a penalty. Bad decision.	Hansen thinks that it was a definite hand-ball.
			Another hand-ball appeal by Stoke. Was hard to see what was going on in real time. The replay confirms that the Southampton player did move his hand toward the ball and it struck his arm. Should have been a penalty.	Hansen also thinks this was a penalty and that Stoke was hard-done by.

Table A3.2 (continued)

Date	Score	Referee	The Play	Commentators' Discussion
Dec. 29, 2012	Sunderland 1 vs. Tottenham 2	Martin Atkinson	Gareth Bale is booked for diving. Looks like a dive in real time. Replay shows that, while Bale goes down a little theatrically, his knee is clipped as he is running at high speed. Should be a penalty and no yellow card.	Hansen thinks the referee got it dramatically wrong. There's definitely contact.
Jan. 1, 2013	Swansea 2 vs. Aston Villa 2	Mark Halsey	Hand-ball by Aston Villa defender, Penny. Referee gives foul right on edge of the box, but it looks inside. From the replay it would appear that the ball is just in the box. Should have been a penalty to Swansea.	For Dublin, it's too close to tell and it's a hard decision for the referee. Probably just inside though.
Jan. 1, 2013	West Ham 2 vs. Norwich City 1	Mark Clattenburg	Penalty claim for West Ham. A Norwich City defender appears to be holding Andy Carroll. Replay shows he is holding him and tugging shirt. Does look like a foul.	Dublin feels this was a definite penalty.
			Kane goes down in the West Ham penalty area. Looked like a penalty in real time. Replay confirms it was a penalty.	Dublin thinks it's a definite penalty. The referee got that decision completely wrong. Lawrenson adds: "The thing is the referees only get one shot at making the decisions." Lineker adds: "It's a hard job."

Table A3.2 (continued)

Date	Score	Referee	The Play	Commentators' Discussion
Jan. 12, 2013	Stoke 0 vs. Chelsea 4	Andre Marriner	Robert Huth brings down Juan Mata of Chelsea in the penalty area. Referee awards the penalty. It looked like a penalty in real time. Looked soft after watching the replay.	Shearer doesn't think it was a penalty.
Jan. 12, 2013	Aton Villa 0 vs. Southampton 1	Mark Halsey	Rodriguez goes down in the box. It looks like a penalty in real time. Replay shows that Rodriguez dived. Bad decision.	From the referee's angle it looks like there's contact, but Hansen doesn't think there's contact and it's definitely a dive. Shearer thinks there should be a retrospective ban. When a dive is like that, someone has to look at that and say it is a dive. Shearer says on first view he thought it was a penalty. Everyone agrees with this. Referees have a difficult job, because they only have one viewing.

Table A3.2 (continued)

Date	Score	Referee	The Play	Commentators' Discussion
Jan. 1, 2013	Fulham 1 vs. Wigan 1	Mark Clattenburg	Damien Duff is brought down in Wigan half. No free kick given. Wigan gets free kick, takes it quick and scores. Bad decision from referee as Duff was clearly fouled.	Shearer feels it was a foul on Duff. Free kick wasn't taken from the proper position and ball was moving.
Jan. 1, 2013	Arsenal 0 vs. Manchester City 2	Mike Dean	Vincent Kompany is sent off for Manchester City. Looked harsh in real time. Referee had great view of the incident. Replay showed it looked like his feet were off the floor. I thought he won the ball, and although it was tough challenge it was never a red card.	Hansen thinks this was nonsense and that this was a perfect challenge. He can't see for the life of him why he was sent off for this.
Jan. 20, 2013	Tottenham 1 vs. Manchester Utd 1	Chris Foy	Wayne Rooney goes down in the box and appeals for a penalty. Looked like a foul in real time, and Rooney was livid afterward. Replay showed Caulker didn't get the ball, and it should have been a penalty.	Hansen thinks it's about as clear-cut as you're ever going to get. Hansen says if they get this it's game over. Player takes the leg. Pat Nevin thinks from the referee's view he only sees that Rooney has kicked his leg out at him. It's up to the assistant referee to give it, because he has the better view. Colin Murray calls it a stonewall penalty.

Table A3.2 (continued)

Date	Score	Referee	The Play	Commentators' Discussion
Jan. 20, 2013	Chelsea 2 vs. Arsenal 1	Martin Atkinson	Arsenal complains that there is a foul on Coquelin by Chelsea's Ramires in the build-up to Chelsea's first goal. Looked like a foul in real time. Replay confirms Arsenal should have had a free kick.	Hansen feels this was a clear foul. Nevin thinks it was hard to see in real time, but it should be a foul.
			Ramires is brought down by Szczęsny for a penalty to Chelsea. Looked like keeper caught him in real time. After watching the replay, it looks like Arsenal is hard-done by as it looks like Ramires actually slips. Poor decision.	Nevin highlights how Ramires slips, so it shouldn't have been a penalty.
Jan. 30, 2013	Fulham 3 vs West Ham Utd 1	Chris Foy	Berbatov scores for Fulham. West Ham appeal for offside. Replay shows he appeared to be offside. Bad decision. The goal shouldn't have stood.	Contentious goal for Shearer. He feels there were three Fulham players offside—Berbatov included.
Feb. 3, 2013	Manchester City 2 vs. Liverpool 2	Anthony Taylor	Džeko goes down under a tough challenge from Liverpool's Škrtel. He stays down, and Manchester City are livid. They gesture for Liverpool to kick it out on several occasions. Liverpool keeps playing and 40 seconds later Sturridge scores. He did look hurt, but I think Škrtel got the ball.	Savage thinks that's a foul.

Table A3.2 (continued)

Date	Score	Referee	The Play	Commentators' Discussion
Feb. 9, 2013	Tottenham 2 vs. Newcastle Utd 1	Phil Dowd	Cabaye only receives a booking for a leg-breaker of a challenge. Late, over the top of the ball. He deserved to be sent off.	Lawrenson thinks it was a very bad challenge and he could have done more damage than he did. If the referee sees that, he's off. Shearer thinks Cabaye was lucky to not get a red card.
Feb. 9, 2013	Stoke 2 vs. Reading 1	Michael Oliver	Stoke's Whelan goes in on a crunching challenge. He should have been sent off. Dangerous challenge.	Lawrenson thinks he was lucky not to be sent off. He also notes that we've seen Whelan do that sort of challenge before.
			Reading striker is brought down in the box. It looked soft in real time, but after looking at the replay he was brought down by a clumsy challenge.	Lawrenson thinks he has a point. It's a foul anywhere else on the field, but the striker goes down a little bit clumsily. Shearer agrees it's a penalty, though the striker went down a bit too easy.
Feb. 23, 2013	Reading 0 vs. Wigan 3	Phil Dowd	Koné scores for Wigan after a scruffy goal from a corner. Reading appeal for hand-ball by Caldwell whose initial header came off Koné for the goal. Was difficult to tell.	Shearer thinks it's a hand-ball by Caldwell. Lineker agrees that it was a hand-ball.

Table A3.2 (continued)

Date	Score	Referee	The Play	Commentators' Discussion
Feb. 23, 2013	West Bromwich Albion 2 vs. Sunderland 1	Roger East	Adam Johnson strikes toward goal but it is blocked by West Bromwich Albion defender. Sunderland appeal for hand-ball. No penalty given. Replay confirms it did hit arm, but looked like ball struck his arm by his side. He did appear to move hand toward ball though.	Hansen thinks it was a penalty. Shearer agrees it was a penalty.
Feb. 23, 2013	Fulham 1 vs. Stoke 0	Lee Probert	N'Zonzi slaps Ruiz in the face as he gets away from him. It was an aggressive slap and he should have been red-carded. Bad decision from the referee just to book him.	Shearer thinks he should have been sent off.
Feb. 24, 2013	Newcastle Utd 4 vs. Southampton 2	Chris Foy	Cissé scores a wonderful goal for Newcastle. Replay subsequently shows that he was offside. Goal shouldn't have counted. Poor decision by the assistant referee.	Shearer says technically he is four yards offside. But if he gives it he wouldn't make it out of the stadium alive. Nevin thinks the assistant referee probably thinks it's touched someone. If you used technology you could clear these things up.

Table A3.2 (continued)

Date	Score	Referee	The Play	Commentators' Discussion
Mar. 2, 2013	Sunderland 2 vs. Fulham 2	Mark Halsey	Philippe Senderos of Fulham pulls the shirt of the Sunderland striker Danny Graham. Penalty given. Replay confirms it was a poor decision. Minimal contact.	Shearer doesn't think that's a penalty at all. Totally agrees with Martin Jol. Hands on, but it's soft.
Mar. 9, 2013	Norwich City 0 vs. Southampton 0	Mark Clattenburg	Holt is pulled back as he tries to slide the ball into the empty net for Norwich City. Looks like a penalty in real time. Replay confirms he was tugged back.	Lawrenson feels this should have been given as a penalty. Shaw hauls him back. Hansen agrees that it was a penalty.
			Holt goes down in the penalty box again. This time, after much deliberation with his assistant referee, Mark Clattenburg gives a penalty to Norwich City. It looked soft in real time. After watching the replay, it looks just as soft. Bad decision.	Lawrenson doesn't think this was a penalty. Assistant referee didn't even flag. Holt went down like a giant redwood. That takes some doing. Hansen similarly thinks it wasn't a penalty. Holt actually pulls back Shaw, and there's minimal contact.

Table A3.2 (continued)

Date	Score	Referee	The Play	Commentators' Discussion
Mar. 9, 2013	West Bromwich Albion 2 vs. Swansea 1	Lee Mason	Swansea goal is chopped off for offside. In real time it looked a harsh decision. After watching the replay it's still confusing as to whether he is offside, because after taking the initial shot, the ball appears to come back to him from a West Bromwich Albion player. I think it's a bad decision.	Hansen thinks it's a goal, because it comes off two West Bromwich Albion players. Hansen says there's no interpretation there; it's about rules. Lawrenson thinks assistant referee gave the decision. If he did, you need to be sure, and how can you be sure? It's not offside.
Mar. 10, 2013	Newcastle Utd 2 vs. Stoke City 1	Andre Marriner	Stoke's Shotton goes down in the box and appeals for a penalty. In real time it looked like Stoke had a good shout. Replay shows there was contact. It should have been given.	Nevin says he's seen them given. Jason Roberts has seen them given.
Mar. 16, 2013	Everton 2 vs. Manchester City 0	Lee Probert	Everton player Kevin Mirallas scores after Osman headers a through ball. Goal is chopped off for offside. Replay shows he was in line and the goal should have stood. Bad decision.	Dublin thinks the ref got it wrong and the goal should have stood.

Table A3.2 (continued)

Date	Score	Referee	The Play	Commentators' Discussion
			Fellaini handles ball. Free kick is given just outside the box. Too hard to tell in real time whether it was in box or not, given the angle. The replay shows Fellaini was about three yards inside the box. Really poor decision. Clear penalty.	Dublin highlights that he was three yards inside the box.
Mar. 17, 2013	Wigan 2 vs. Newcastle Utd 1	Mark Halsey	Wigan's McManaman goes in with a shocking challenge on Newcastle player. High (above the knee) and late. Shocking challenge, which should have resulted in a red card. Poor decision to give nothing. Player was stretchered off. Don't think referee saw it clearly enough.	Lawrenson says the challenge is horrendous. Straight red. Lawrenson says that Halsey just didn't see it. It looks like the assistant referee should see it. Nevin says it's an awful challenge. One of the worst he's seen.
			Wigan score amidst appeals that there was a hand-ball when cross came over. Looked clear at the time. Replay confirms it was a clear hand-ball. Should have been disallowed.	Nevin says it's a clear hand-ball, but they didn't notice it until watching replay. Lawrenson says its shades of Maradona.
Mar. 17, 2013	Sunderland 1 vs. Norwich City 1	Chris Foy	Norwich City appeal for hand-ball after cross comes off Danny Rose. Looked like hand-ball. Referee gives free kick on edge of the box. Replay shows he was inside the box. Poor decision.	Lawrenson thinks this was a penalty. Nevin says this is as stonewall penalty and again stresses about the use of video technology.

Table A3.2 (continued)

Date	Score	Referee	The Play	Commentators' Discussion
Mar. 30, 2013	Arsenal 4 vs. Reading 1	Chris Foy	Giroud knocks the ball over the keeper, but is then taken out the game. It looks like a penalty in real time and replay confirms he was absolutely clattered by the goalkeeper Stuart Taylor. Should have been a penalty.	Shearer says, "How that is not a penalty I do not know. How the referee doesn't see it I've no idea."
Mar. 31, 2013	Aston Villa 1 vs. Liverpool 2	Lee Mason	Villa's Westwood catches Jordan Henderson right above the knee. Shocking challenge. Should have been a red card.	Savage does think this was a red card. Steven Reid says that's a career-ending challenge. Red card.
Apr. 6, 2013	Stoke 1 vs Aston Villa 3	Mark Clattenburg	Jonathan Walters crosses the ball into the box, but it strikes the arm of the Aston Villa defender. Looks soft in real time. Replay confirms it hit defender's arm and it was in an unnatural position. Bad decision.	Lineker says the rules are that it needs to be deliberate, but that is fairly ambiguous. Penalty.
Apr. 6, 2013	Norwich City 2 vs. Swansea 2	Michael Oliver	Grant Holt should have been sent off for a bad tackle. It was a stamping action and he deserved to go. Not even a yellow card for the challenge. Bad decision.	Shearer thinks he should have got a red card for this challenge. Studs down his leg.

Table A3.2 (continued)

Date	Score	Referee	The Play	Commentators' Discussion
Apr. 7, 2013	Liverpool 0 vs. West Ham Utd 0	Anthony Taylor	West Ham appeal for penalty, as Tomkins appears to be kicked by Enrique in the box. Replay shows there was contact, and I've seen them given.	Hansen thinks Enrique misses this and kicks Tomkins. He thinks it's a penalty. Kevin Kilbane also thinks this is a clear penalty.
Apr. 13, 2013	Arsenal 3 vs. Norwich City 1	Mike Jones	Olivier Giroud of Arsenal is brought down in the box. Doesn't look like much at the time and didn't seem to be many appeals from Arsenal players (including Giroud himself). Assistant referee gives the penalty rather than the referee. Looks like a tough decision against Norwich City after viewing the replay, although his shirt was pulled back from Bassong who was on the ground.	David Moyes doesn't think it was a penalty and it was harsh on Norwich City. Hansen doesn't think it was a penalty, and thinks there was a debate as to whether it was a corner kick. Massive call for a linesman 45 yards away, looking through a crowd of players.
			Podolski scores third for Arsenal. Replay shows that Walcott was offside in the build-up. Goal shouldn't have stood. Bad decision. Giroud also fouled the defender in the build-up.	Keown thinks it was a foul by Giroud and that Walcott was offside. Technology line confirms it.
Apr. 14, 2013	Newcastle Utd 0 vs. Sunderland 3	Howard Webb	Cissé scores for Newcastle, but it's chopped off for offside. It looked tight at the time. Replay shows he was onside. Really poor decision by the assistant referee.	Lawrenson thinks he is most definitely on. Lawrenson says you can't guess; you can only give it if you're 100% sure. Roberts also thinks this was onside.

Table A3.2 (continued)

Date	Score	Referee	The Play	Commentators' Discussion
Apr. 20, 2013	Sunderland 1 vs. Everton 0	Phil Dowd	Jelavić goes down in the box and appeals for a penalty for Everton. Looked like he had a shout. Replay shoes Danny Rose was pulling him back. Should have been a penalty.	Hansen thinks he's holding him, and dragging him. It's a penalty. Shearer thinks this wasn't a penalty. If he wanted to head it, he'd head it. Given that there's a split, I'm going with my own judgment on the incident, which is that it was a penalty.
Apr. 20, 2013	Swansea 0 vs. Southampton 1	Mark Halsey	Southampton scores, but it's chopped off for a foul/hand-ball by Lallana on the keeper. Assistant referee gives the foul and I think it's a really harsh decision. Should have been a goal.	Shearer has watched the incident a half dozen times, and he can't see why he's given it. He doesn't think it hits his hand at all. He thinks Southampton is hard-done by.
Apr. 21, 2013	Liverpool 2 vs. Chelsea 2	Kevin Friend	Luis Suarez bites Ivanović. Referee didn't see it. That should have been a sending off.	Savage is in disbelief. Red mist has come over. It shouldn't happen. Hansen says it's appalling. Massive flaw in character.

Table A3.2 (continued)

Date	Score	Referee	The Play	Commentators' Discussion
Apr. 27, 2013	Wigan 2 vs. Tottenham 2	Martin Atkinson	Kyle Walker goes down under a challenge just outside the box. Looks like both men are jostling and trying to get to the ball. Decision could have gone either way, but Walker actually looks like he pulls back the defender. Foul went to Spurs. Spurs scored directly from the free kick to equalise.	Shearer hinks it was a foul for Wigan. Walker's hands are all over the Wigan defender. He thinks they were harshly treated.
May 4, 2013	West Bromwich Albion 2 vs. Wigan 3	Lee Probert	West Bromwich Albion player crosses into the box strikes the Wigan defender's arm. Looks unclear in real time how much intent there was. Replay confirms his arm was raised and it should have been a penalty to West Bromwich Albion.	Hansen thinks it's a gray area. Looks to him he's handled it. Very lucky to get away with it; on another day it's given.
			West Bromwich Albion's Jones gets pushed down in the box by Wigan's Thomas. Looks soft in real time. Replay confirms the player was pushed out of the way of the trajectory of the ball. Should have been a penalty.	Hansen thinks that's a penalty all day long. He knows what he's doing and he blocks him off. Shearer's not sure that's a penalty. He thinks the defender stands his ground. Because they are split, I've gone with my own judgment, which is that it was a penalty.

Table A3.2 (continued)

Date	Score	Referee	The Play	Commentators' Discussion
May 4, 2013	Norwich City 1 vs. Aston Villa 2	Kevin Friend	Bennett pulls down Norwich City's Snodgrass for a penalty. He did have arm around him and he brought him down. Bennett should have received a second yellow card.	Hansen thinks Hughton has a point and he doesn't whine normally. Bennett commits several bookable offenses in the game. He is exceptionally lucky to stay on the pitch.
May 4, 2013	Swansea 0 vs. Manchester City 0	Mike Jones	Michu is brought down in the Manchester City box. Looks soft in real time. Replay shows that he was clattered in the box, and the defender got nowhere near the ball. Stonewall penalty. Poor decision.	Lineker thinks this should have been a penalty.
			Džeko goes down in the Swansea box under a challenge from Chico Flores. In real time I think it looks like a strong penalty shout. Replay shows that he was looking for Flores's leg which was out but then there was contact.	Lineker thinks this should have been a penalty. Hansen thinks there were a couple of decent penalty appeals in the game.
May 5, 2013	Liverpool 0 vs. Everton 0	Michael Oliver	Distin scored for Everton, but the referee disallows it for a foul by Anichebe on the Liverpool goalkeeper. Replay showed there wasn't much in it. Everton should feel very hard-done by.	Nevin thinks there's nothing wrong with this. He is standing his ground. Shearer says Anichebe shouldn't have to get out of the way for the keeper. He doesn't see anything wrong with the goal.

Table A3.2 (continued)

Date	Score	Referee	The Play	Commentators' Discussion
May 11, 2013	Aston Villa 1 vs. Chelsea 2	Lee Mason	Aston Villa's Benteke swings elbow at Azpilicueta. It looked like a sore one in real time. Replay showed it was very bad and he's lucky not to have been sent off. He received a yellow card only.	Dublin says this is the bad part of Benteke. When Dublin watched it from stands, he was thinking: "What are you doing? Get up, he hasn't touched you." But after seeing it now why do you need to do that? Savage thinks this is definitely a red card. 100%
May 19, 2013	Tottenham 1 vs. Sunderland 0	Andre Marriner	Gareth Bale is brought down at pace by Larsson in the Sunderland box. Looks like a stonewall penalty in real-time, but Gareth Bale is booked for simulation. Bale is livid. Replay shows Larsson had hands all over him pulling him back. Shocking decision (It's worth noting that Bale had been booked for diving three or four times that season, so perhaps his reputation preceded him.) Certain penalty.	Hansen for the life of him can't understand why the referee has booked him for diving. The referee was a long way off and he's speculating. Shearer thinks the referee's position it's terrible. 1. He's got to be sure it's a penalty. 2. He's got to be sure if he's diving. From the distance he's away there's no way he can tell. It's not a dive; it's a clear-cut penalty.

Table A3.2 (continued)

Date	Score	Referee	The Play	Commentators' Discussion
May 19, 2013	Liverpool 1 vs. Queens Park Rangers 0	Martin Atkinson	Coutinho headers the ball toward Queens Park Rangers' goal. He appeals that ball crosses line, but it's cleared away and no goal is given. Looked close in real time. Replay shows it was over. Goal should have been given. Commentator said: "Goal-line technology can't come too soon."	Pundits didn't discuss, but given how much the ball was over the line I've included it in my analysis here.

Table A3.3

Wrong refereeing decisions: Season 2013–2014

Date	Score	Referee	The Play	Commentators' Discussion
Aug. 24, 2013	Hull City 1 vs. Norwich City 0	Michael Jones	Penalty to Hull City as Sagbo is pushed in box by defender, Turner. Looked like penalty in real time. Replay shows there was minimal contact and there would be penalties all the time if that is a penalty.	Alan Hansen doesn't think it's a penalty.
			Turner of Norwich City goes down in the Hull City box and claims for a penalty. Nothing given. Looked like a shove in real time, but there was no replay as subsequent incident overshadowed it.	Hansen thinks this is definitely a penalty.

Table A3.3 (continued)

Date	Score	Referee	The Play	Commentators' Discussion
Aug. 26, 2013	Tottenham 1 vs. Swansea 0	Neil Swarbrick	Tottenham's Townsend is brought down by Shelvey on the edge of the box. Referee gives a free-kick, amidst protests from Spurs players he was inside. Replay shows he was inside and it was a stonewall penalty. Poor decision.	Alan Shearer thinks he got the first one wrong. His positioning was good, but the assistant needed to help the referee more with this decision. It's a stonewall penalty.
			Townsend goes down in the box, but this time a penalty is given against Shelvey. There appeared to be less contact than before, but he didn't complain. Was touch-and-go, but he probably got the decision right.	Shearer doesn't think this is a penalty. He thinks because the referee was under so much pressure for the first bad decision, that he gives this one.
Aug. 31, 2013	Norwich City 1 vs. Southampton 0	Howard Webb	Southampton's Lallana strikes the ball towards goal and claims hand-ball. Player looks too close to the ball. Replay shows player's hands were up high and it should have been a penalty. Poor decision.	Robbie Savage thinks this was a blatant hand-ball. He has no idea why Webb hasn't seen it; his position is perfect. Michael Owen doesn't think it's deliberate, but it's a definite hand-ball. If you put your hands up that high, it's a penalty.

Table A3.3 (continued)

Date	Score	Referee	The Play	Commentators' Discussion
Aug. 31, 2013	Cardiff 0 vs. Everton 0	Anthony Taylor	Leighton Baines goes down in the box and it looks like a penalty. Replay shows it should have been a penalty. Referee doesn't book him for diving so not sure why he didn't give it. Poor decision.	Savage thinks he was clever because he collapses his leg, but thinks it's a definite penalty.
Sept. 14, 2013	Sunderland 1 vs. Arsenal 3	Martin Atkinson	Sunderland's Altidore is blatantly held by Sagna, but he carries on and scores. Referee doesn't play the advantage for some reason and gives the foul instead. Poor decision.	Shearer feels the referee had a shocker and has to wait in order to try and play the advantage.
Sept 14., 2013	Manchester Utd 2 vs. Crystal Palace 0	Jon Moss	Ashley Young goes down and penalty is given. In real time it looked like Young was already off-balance and falling—played for it. Referee took a long time to give the decision. It looked harsh after viewing the replay and in actual fact was outside the box. Player was sent off also for denying a goal-scoring opportunity.	Shearer doesn't think it's a penalty. Referee looks to assistant, but doesn't really get help. He's 45 yards away and there's no way he can be certain about it being in the box, or being a goal scoring opportunity. He got it absolutely wrong.

Table A3.3 (continued)

Date	Score	Referee	The Play	Commentators' Discussion
Sept. 15, 2013	Southampton 0 vs. West Ham Utd Utd 0	Andre Marriner	Southampton's Schneiderlin goes in two-footed on Diamé. Looks like a bad challenge. Feet were low, but it was two-footed and he did catch the ankle of Diamé. Could have been a red card.	Savage thought it was a horrible challenge and thinks it's a red card all day long. He thinks referee is in a good position and he has to give it. That kind of challenge could break someone's leg. Mark Lawrenson can't understand why the referee hasn't given it.
			Appeals for hand-ball as West Ham Utd's Kevin Nolan flicks ball with arm. In fairness to him he was trying to move his arm away, but it did strike his arm. I think it should have been a penalty.	Lawrenson thinks they have a massive case for a penalty. Nolan has got away with one here. It's a penalty. Savage also thinks it was a penalty.
Sept. 21, 2013	Liverpool 0 vs. Southampton 1	Neil Swarbrick	Daniel Sturridge goes down under a challenge from Lovren. Looked like a penalty in real time and replay shows he got nothing on the ball. Poor decision.	Fowler's first impression was that it was a penalty. Upon looking at the replay he thinks it was. Feels Sturridge was too honest. Hansen thinks it was a penalty.

Table A3.3 (continued)

Date	Score	Referee	The Play	Commentators' Discussion
Sept. 28, 2013	Hull City 1 vs. West Ham Utd 0	Kevin Friend	Hull City's Brady goes down in the box and penalty is given. Looked soft in real time and replay confirms he went down theatrically under little contact from the West Ham Utd defender O'Brien. Contentious decision.	Danny Murphy doesn't think it is a penalty. He felt there was minimal contact, but you'd be disappointed to have that given against you. Roberto Martinez would be disappointed to concede a penalty like that. Contact doesn't always mean a penalty.
			West Ham Utd appeal for a hand-ball in the penalty area. Looked like there was contact in real time and replay confirms it was a definite penalty as defender moved arm out towards the ball. Poor decision not to spot it by the referee.	Murphy thinks it's a clear penalty. He thinks the referee's line of sight is a little bit difficult, players in between him and the ball. But it's a penalty.
Sept. 29, 2013	Sunderland 1 vs. Liverpool 3	Howard Webb	Sturridge scores for Liverpool, with a header from a corner. Looked good in real time. Replay show Sturbridge flicked it in with his arm. Goal should have been disallowed. Poor decision.	Peter Schmeichel says the goal shouldn't have stood.

Table A3.3 (continued)

Date	Score	Referee	The Play	Commentators' Discussion
Sept. 29, 2013	Stoke City 0 vs. Norwich City 1	Anthony Taylor	Stoke's Kenwyne Jones is pulled down in the box. Looked like a foul in real time and replay confirms he was grabbed by Norwich City midfielder, Fer. Although he went down theatrically, he was pulled back.	Jason Roberts shows referee's position is blocked. Roberts says there is contact, Jones accentuates it but that is not the same as diving. What the player is doing is making sure that the referee can see he is being held. Absolute penalty for Roberts. Schmeichel agrees with Roberts. He thinks it's a stonewall penalty. Schmeichel also mentions that Fer had already been booked and thus would have been sent off.
Oct. 5, 2013	Manchester City 3 vs. Everton 1	Jon Moss	Manchester City's Nastasić pushes Lukaku in the box. Was a clear penalty. Nothing given. Poor decision.	Shearer agrees with Martinez that this was a penalty. It's a clear shove.
Oct. 5, 2013	Fulham 1 vs. Stoke 0	Roger East	Walters of Stoke is brought down in the box. Free kick is given, but the main foul happened well inside the box. Strange and poor decision. Should have been a penalty.	Shearer thinks this is a penalty.
			Fulham's Bryan Ruiz goes down in the box. No penalty given. There was contact, although I felt he went down too easy. Touch-and-go.	Shearer thinks this is also a penalty, and that the referee makes the wrong decision.

Table A3.3 (continued)

Date	Score	Referee	The Play	Commentators' Discussion
Oct. 6, 2013	West Bromwich Albion 1 vs. Arsenal 1	Lee Mason	Laurent Koscielny jumps up for the ball and pushes West Bromwich Albion's Shane Long down in the box. It looked like a penalty in real time and replay confirms it. Bad decision.	Hansen thinks this was a penalty, all day long. John Hartson also thinks it's definitely a penalty.
Oct. 6, 2013	Norwich City 1 vs. Chelsea 3	Neil Swarbrick	Norwich City's Pilkington goes down under the challenge of Ramires. Looked like a penalty in real time. Replay shows Chelsea player got nothing on the ball. Poor decision.	Hansen thinks this was a penalty kick. He doesn't get anywhere near the ball, so it's a poor decision.
Oct. 19, 2013	Chelsea 4 vs. Cardiff 1	Anthony Taylor	Eto'o kicks the ball out of Marshall's hands as he bounces it before kicking it out. He wasn't in control of the ball anymore, but it's a tough call by the referee. Was he still in control of the ball when he drops it in that fashion?	Shearer says the law is simple. Even though the keeper is bouncing the ball he is still deemed to have it under control, so it's a free kick. Why on earth they haven't seen that is beyond him.

Table A3.3 (continued)

Date	Score	Referee	The Play	Commentators' Discussion
Oct. 26, 2013	Manchester Utd 3 vs. Stoke 2	Lee Mason	Javier Hernández catches Stoke defender Robert Huth. Stoke appeal for red card. Challenge looked late and high in real time. Have seen people sent-off for challenges like that.	Hansen feels on another day it might have been red.
Oct. 26, 2013	Crystal Palace 0 vs. Arsenal 2	Chris Foy	Arsenal full-back Serge Gnabry goes down on the edge of the box amidst appeals for a penalty. Looked very tight in real time. Referee gives free kick on the edge of the box, but replay shows it was on the line and probably should have been a penalty. Contentious decision.	Savage thinks this was a definite penalty. It's in the box.
Oct. 26, 2013	Aton Villa 0 vs. Everton 2	Anthony Taylor	Everton's James McCarthy went in late and high on Villa player. Looked bad in real time. Replay suggests it could have been a red. I've seen them given. He only received a yellow card.	Lineker thinks he was lucky to be on the field. Savage thinks this was a red card. Referee had a good view of it, but deemed it a yellow. Poor decision. Hansen also thinks it was a red card.

Table A3.3 (continued)

Date	Score	Referee	The Play	Commentators' Discussion
Oct. 27, 2013	Tottenham 1 vs. Hull City 0	Michael Oliver	Hull City's El Ahmadi is penalized for hand-ball. Penalty to Tottenham. Looked like hand-ball at the time and replay confirmed his arm was slightly raised— although he was close to the ball and there was no intent. Harsh decision, given the distance from the ball, but his arm was raised.	Lawrenson thinks it's a guess from the referee, as he can't see the incident and can't possibly see his hand hit the ball. He questions whether it would have been given at the other end of the field. Owen thinks it was a joke. He's a big fan of Michael Oliver, but that's a poor decision and Steve Bruce has every right to be upset. His hands aren't above his head, and the ball flicks up his thigh. He then shows a camera angle which suggests that the referee can't really see the incident.
Nov. 2, 2013	Fulham 1 vs. Manchester Utd 3	Lee Probert	Fulham defender Sascha Riether stamps on Januzaj in stoppage time. He was nowhere near the ball—it's a clear red card. Referee didn't give anything. He was very fortunate, as it should have been a red card.	Murphy thinks he's a lucky boy. He should have been sent off.

Table A3.3 (continued)

Date	Score	Referee	The Play	Commentators' Discussion
Nov. 2, 2013	Hull City 1 vs. Sunderland 0	Andre Marriner	Hull City's Brady jumps in on Allan Johnston. Sunderland appeal for red card, but only a yellow given. Looked like a bad foul in real time, and replay confirms this. I don't think it was a red card necessarily, but both feet were off the ground and by the letter of the law he had to go.	Shearer feels this was dangerous and he should have been sent-off. Murphy thinks by the letter of the law this should have been a red card.
Nov. 11, 2013	Chelsea 2 vs. West Bromwich Albion 2	Andre Mariner	Ramires goes down for Chelsea and penalty is given late in the game. Looked a bit soft in real time and West Bromwich Albion is furious. Replay shows it was shoulder-to-shoulder, and if anything Ramires drifted his body toward the defender to bring on the contact. Poor decision.	Shearer thinks this is a shocking decision. Terrible decision. Ramires runs into him. Owen agrees and thinks he plays for it and goes down very easy.

Table A3.3 (continued)

Date	Score	Referee	The Play	Commentators' Discussion
Nov. 10, 2013	Swansea 3 vs. Stoke City 3	Robert Madley	Stoke is awarded a penalty for hand-ball. It didn't look like a hand-ball in real time. Replay showed ball barely glanced off arm of Routledge after deflection. Referee had a poor view, so I've no idea what he thinks he saw. Really poor decision.	Keown thinks there's absolutely no way that that's a penalty. Ball hits him, but how can he get out of the way of that? Shearer thinks the referee has a poor view, and there's absolutely no way he can give that. There's no way that should be a penalty. Shearer had to watch it three or four times, so at normal speed there's no way referee can say that's it a hand-ball.
Nov. 10, 2013	Sunderland 1 vs. Manchester City 0	Mike Dean	Sunderland's Sebastian Larsson commits a bad foul on City player. It was over the ball, late, and he caught the player high on the shin. Although it didn't seem malicious, it should have been a red card.	Keown thinks this should have been a red card. He got away with it. It's a leg-breaker. Shearer thinks the referee had a good view of it and got it completely wrong. It should have been a red card.
			Phil Bardsley scores for Sunderland amidst appeals he pushed the City defender. Looked like a foul after looking at the replay.	Keown thinks it was a push by Bardsley and a foul. Shearer disagrees and thinks Bardsley was stronger. Given the pundits were split on this decision, I went with my own view which is that this was a foul.

Table A3.3 (continued)

Date	Score	Referee	The Play	Commentators' Discussion
Nov. 23, 20/13	Everton 3 vs. Liverpool 3	Phil Dowd	Everton's Mirallas catches Luis Suárez. It looks high and dangerous. Should be a red card as he catches him knee high. The incident was in full view of Phil Dowd. Referee booked him.	Hansen thinks he is very lucky to be on the pitch. From the first angle it's bad but from the second angle it's terrible. Kevin Kilbane agrees that he should have been sent off.
Nov. 23, 2013	Stoke 2 vs. Sunderland 0	Kevin Friend	Wes Brown of Sunderland is sent off for a flying tackle on Charlie Adam. Looked dangerous in real time. Sunderland manager Poyet can't believe the decision. Replay shows that while it could be said to be dangerous play, he did clearly play the ball. Poor decision.	Hansen fully agrees with Gus Poyet. It's not a sending off. It's not high; it's not over the top. There's a degree of aggression in the follow through, but are you meant to tackle without aggression? It's a joke, he gets 100% of the ball, and the referee got it totally wrong. Kilbane can't understand the decision. He says the referee takes a few seconds to make the decision, but gets it wrong.

Table A3.3 (continued)

Date	Score	Referee	The Play	Commentators' Discussion
Nov. 24, 2013	Cardiff 2 vs. Manchester Utd 2	Neil Swarbrick	Wayne Rooney tussles with Cardiff's Mutch, and then takes a kick at him off the ball. Should have been a red card, but Rooney was only booked.	Shearer felt the referee didn't deal with this in the right way. It was a bad tackle and it should have been a red card. He kicks out at him. If the referee saw him kick out (he gave a yellow card) then why didn't he send him off?
Dec. 4, 2013	Swansea 3 vs. Newcastle Utd 0	Howard Webb	Newcastle's Loïc Rémy's shot is blocked on way to Swansea goal. He appeals for hand-ball. Looked like a hand-ball in real time. Replay shows it was a hand-ball and it should have been a penalty. Poor decision.	Shearer thinks his hand was out and it's a definite penalty. Robbie Fowler thinks this was a penalty.
			Shelvey makes headbutt gesture towards Debuchy. It's not much contact, but he does motion toward him and make contact. Should have been a red.	Shearer thinks this was a red card. Fowler thinks this could have been a red.

Table A3.3 (continued)

Date	Score	Referee	The Play	Commentators' Discussion
Dec. 4, 2013	Stoke City 0 vs. Cardiff 0	Michael Oliver	Charlie Adam pushes Fraizer Campbell onto the ground during a fracas in the wall before a free kick. Adam had already been booked and should have been sent off.	Shearer thinks he should have been sent off. It's a case of mistaken identity, as the referee thinks its Peter Crouch who has done it and he books him. Fowler thinks it could have been a straight red card.
Dec. 12, 2013	Stoke 3 vs. Chelsea 2	Jonathan Moss	Jonathan Walters appears to foul Azpilicueta on halfway line, before setting up Stephen Ireland for Stoke's second goal. Replay shows there was contact between the pair and I have seen fouls given for less.	Savage thinks it's a foul by Jonathan Walters after looking at the replay.
Dec. 7, 2013	Liverpool 4 vs. West Ham Utd 1	Michael Oliver	Flanagan commits a bad challenge on Downing. Foul given, but should have been a yellow if not red.	Shearer thinks Flanagan should have been sent off and was lucky because it was the third minute of the game.
Dec. 7, 2013	Sunderland 1 vs. Tottenham 2	Lee Mason	John O'Shea claims there was a hand-ball in the box by a Spurs defender. The replay shows that the ball did strike Sandro's arm and that his arm was high in the air. It should have been a penalty.	Savage thinks it should have been a penalty, no doubt about it. Real time, slow motion, however you look at it, he thought it was a penalty. Can't believe ref (who saw it directly—they showed his line of sight) didn't give it.

Table A3.3 (continued)

Date	Score	Referee	The Play	Commentators' Discussion
Dec. 8, 2013	Fulham 2 vs. Aston Villa 0	Mike Dean	Kačaniklić of Fulham is brought down for penalty. Looked soft in real time. Replay shows Bacuna did catch him, although I think the Fulham player backed into him. Soft penalty.	Hansen doesn't think this was a penalty. Player barges into Bacuna and then goes over. Danny Mills agrees that the striker barged into the defender. Poor decision.
			Villa's Agbonlahor goes down in the box and appeals for a penalty. Looked like a penalty in real time. Replay confirms Hughes of Fulham catches him. Poor decision.	Mills thinks this was a penalty. Although player goes down theatrically, replay shows there is definite contact and that's a foul. He had a good view, he was unobstructed.
Dec. 21, 2013	Manchester Utd 3 vs. West Ham Utd 1	Michael Jones	Januzaj goes down in the box under a challenge from Noble. No penalty given. Replay confirms that Januzaj was pushed by Noble and a penalty probably should have been given.	Kilbane thinks there was contact.
Dec. 21, 2013	West Bromwich Albion 1 vs. Hull City 1	Jonathan Moss	Morrison is fouled in build-up to Hull City goal. Replay shows he was fouled directly in front of the referee.	Hansen mentions that he thinks he might have been fouled, but doesn't excuse the West Bromwich Albion defence.

Table A3.3 (continued)

Date	Score	Referee	The Play	Commentators' Discussion
Dec. 23, 2013	Arsenal 0 vs. Chelsea 0	Mike Dean	John Obi Mikel goes in hard on Mikel Arteta on halfway line. Looked like a crunching challenge. Replay shows he was lucky to stay on the park. Should have been a red card.	Keown says there is no care shown for his opponent and he is surprised he wasn't sent off. He thinks it looks particularly bad when you watch it in slow motion, because at match speed it looked like he got the ball. Murphy feels it was late, and he thinks he meant it. He didn't think it was a sending off, although he agrees with what Keown is saying.
			Arsenal's Walcott gets pulled down in the box by Willian. Looked like a penalty in real-time, but referee didn't give it. After watching the replay, it looked like a poor decision.	Murphy doesn't think this was a penalty, from what the referee saw. From that perspective it looked like a nothing coming together of two players. However, he and Keown had the luxury of seeing three or four replays and it looks like a penalty. Keown thinks it should have been a penalty, although Walcott goes down rather easily.

Table A3.3 (continued)

Date	Score	Referee	The Play	Commentators' Discussion
Dec. 26, 2013	Manchester City 2 vs. Liverpool 1	Lee Mason	Raheem Sterling is played in by Suárez and is right through on goal, but he's called offside. Replay shows he was miles onside.	Shearer says it's a terrible, terrible decision. Lineker says it's not even close to being offside.
			Suárez appeals for a penalty. Replay shows his shirt is being held, but commentator notes that he also went down theatrically. I've seen them given, however.	Shearer says while Rodgers is right to bemoan Suárez being held, if referee gave penalty for that then Škrtel would have had to be punished for continually holding also.
Dec. 26, 2013	Newcastle 5 vs. Stoke 1	Martin Atkinson	Glenn Whelan sent off for second yellow card. He and Cabaye had been in argument previously and second challenge produced another yellow card. Looked soft in the replay, but in light of previous argument made sense.	Shearer agrees with Hughes that the first yellow card is harsh. Savage thinks it's harsh.
			Before the cross that leads to Newcastle's second goal, the ball appears to go out of play. Replay confirmed that the ball did go out before Ben Arfa crossed it.	Shearer thinks ball went out, but that the assistant should be giving that, not the referee. Feels a lot of decisions went against Stoke. Savage says the ball went out of play and the goal shouldn't have counted.

Table A3.3 (continued)

Date	Score	Referee	The Play	Commentators' Discussion
Dec. 29, 2013	Chelsea 2 vs. Liverpool 1	Howard Webb	Samuel Eto'o of Chelsea catches Liverpool player above the knee in a bad challenge. Looked bad in real time. Replay shows it was a very bad challenge. I've seen red cards given for that. I think because it was so early in the game, the referee was lenient. Liverpool scored from the resulting free kick.	Shearer thinks this was high, dangerous, and should have been a red. Howard Webb is five yards away; he gets a good view of it.
			Bizarre incident as Eto'o needlessly takes a kick at Suárez in the penalty area, after Chelsea defender had already tackled Suárez and cleared the ball. Should have been a penalty to Liverpool. Referee had a great view of it, so I've no idea why he didn't give it.	Shearer thinks this was a penalty.
Jan. 1, 2014	Manchester Utd 1 vs. Tottenham 2	Howard Webb	Hugo Lloris runs out to challenge Ashley Young and appears to take him out. Replay shows that Young beat him to the ball, and he gets taken out.	Dietmar Hamaan thinks it's a blatant penalty. Reckless challenge. Definite penalty and a booking. Fowler thinks it's a penalty.

Table A3.3 (continued)

Date	Score	Referee	The Play	Commentators' Discussion
Jan. 1, 2014	Sunderland 0 vs. Aston Villa 1	Michael Jones	Giaccherini scores for Sunderland, but he is deemed to be offside. Replay shows he was level. Wrong decision.	Hamaan feels he is level. If there is doubt in linesman's mind, he should keep flag down. It should have been 1–1.
Jan. 1, 2014	Crystal Palace 1 vs. Norwich City 1	Mike Dean	Marouane Chamakh is shown yellow card for tussle with opponent. Upon looking at the replay, it's clear that Chamakh shoved the player hard in the face and should have been red-carded for violent conduct.	Fowler says that by the letter of the law if you push someone in the face you go. It's a blatant red card, and he doesn't know how that can't be given. Lineker agrees it's a red card. Hamaan can't believe that if the referee saw the incident (which he must have if he gave him a yellow), then why didn't he send him off.
Jan. 11, 2014	Fulham 1 vs. Sunderland 4	Mike Dean	Adam Johnson is clipped by Sidwell and scores from the accompanying free kick. Replay shows there was contact, but Johnson went down two steps later.	Malkay Mackay thought there was contact, but Johnson went down a couple of steps later when he realized ball was away from him.
Jan. 11, 2014	Cardiff 0 vs. West Ham Utd 2	Lee Mason	Fraizer Campbell appeals that he is being held in the box. Replay shows that McCartney does have a couple of arms around Campbell, and I felt that should have been a penalty.	Hansen feels the referee might have given that one.

Table A3.3 (continued)

Date	Score	Referee	The Play	Commentators' Discussion
Jan. 12, 2014	Stoke City 3 vs. Liverpool 5	Anthony Taylor	Bad clearance by Stoke player appears to strike the hand of Raheem Sterling who then drives into the box to earn a penalty. Replay showed it was hand-ball. Sterling goes down in the box under a soft challenge. I think it was a soft penalty.	Lawrenson thought this was a hand-ball. His arm is high above head, should have been a foul. Mills also thought this was a hand-ball. Lawrenson doesn't think this was a foul at all. Mills thought it was a soft penalty. From the referee's angle you can see why he's maybe gave it, but even Sterling didn't appeal massively for it.
Jan. 12, 2014	Newcastle 0 vs. Manchester City 2	Mike Jones	Newcastle scores a great goal, but then referee chops off goal for offside. In real time there didn't look like much wrong. Replay shows Newcastle player was standing in an offside position and deemed to be interfering with play. I think it's a poor decision.	Mills feels none of the laws apply to this goal. He wasn't impeding an opponent and he wasn't interfering with the line of sight of the goalkeeper. He thinks the referee has got it wrong. The linesman doesn't raise his flag. Completely wrong decision. Lawrenson thinks there's an element of doubt with the referee: "Have I got this right if I give a goal?"
Jan. 19, 2014	Chelsea 3 vs. Manchester Utd 1	Phil Dowd	Vidić is sent-off for foul on Hazard. Was late and a bad challenge, but I don't think it was a red card.	Hansen, Schmeichel, and Nevin all think this was a yellow card.

Table A3.3 (continued)

Date	Score	Referee	The Play	Commentators' Discussion
			Rafael goes in two-footed and high on Chelsea's Gary Cahill. Stupid, shocking challenge. He gets booked, but it should have been red.	Hansen, Schmeichel, and Nevin all think this should have been a red card.
Jan. 29, 2014	Tottenham 1 vs. Manchester City 5	Andre Marriner	Danny Rose brings down Džeko in the box. Linesman raises the flag to award a penalty to Manchester City. Replay shows Rose got his foot on the ball. He came in from a wrong angle and I think that's why it was given. That said, it shouldn't have been a penalty. Rose is sent off.	Hansen feels it was a big decision that the referee got completely wrong. Referee is a long way off and has bad view on incident. The assistant flags, and it's a bad decision.
Feb. 1, 2014	West Ham Utd 2 vs. Swansea 0	Howard Webb	Andy Carroll is sent off for swinging elbow at Swansea's Chico Flores. Looks bad in real time. Replay makes it look like there isn't an enormous amount to it, in all honesty. Carroll's arm appears to be swung because of his momentum and because of Flores falling down on top of him. Harsh decision I think. Flores made the most of it.	Hansen thinks it's harsh. He definitely comes in with his arm, but it's not deliberate and there's minimal contact. Hansen thinks the FA will rescind the card.

Table A3.3 (continued)

Date	Score	Referee	The Play	Commentators' Discussion
Feb. 8, 2014	Norwich City 0 vs. Manchester City 0	Jonathan Moss	Yaya Touré has scuffle with Norwich City player. Replay shows Touré kicked him in the back. It should have been an instant red card. Referee didn't even give a foul.	Hamaan thinks Touré will be in trouble with the FA after that kick because the rules say if the referee doesn't see the incident they can still act. He feels the player overreacted, but Touré gave him the opportunity to do so.
Feb. 22, 2014	West Ham Utd 3 vs. Southampton 1	Mark Clattenburg	Jarvis scores for West Ham Utd. Replay shows he was a half-yard offside. Commentator thought it was very tight. I thought he was offside.	Hansen (with the help of a video replay line) shows that Jarvis was offside for the goal.
Mar. 1, 2014	Stoke 1 vs. Arsenal 0	Michael Jones	Charlie Adam runs over and stamps on Giroud's leg while he's on the ground. The question is whether it was inadvertent or whether Adam meant it. Knowing Adam, I have a feeling he might have meant it. He should have been sent off. Perhaps not coincidentally, he was taken off two minutes later.	Hansen thinks Adam knows he's there and is very lucky to be on the pitch.

Table A3.3 (continued)

Date	Score	Referee	The Play	Commentators' Discussion
			Hand-ball by Koscielny in the area. Replay seemed to suggest Arsenal was very hard-done by. It was a gangly challenge by the defender, but the ball was played up onto his arm and I don't think there's much he could have done about it.	Hansen thinks it hits his hand, but there's no intent and it's very harsh. Savage thinks it's a penalty. Lineker is surprised Savage thinks that.
Mar. 1, 2014	Everton 1 vs. West Ham Utd 0	Jonathan Moss	Nolan goes down outside the box under a challenge from Gareth Barry. Replay shows Barry had arms around him and he was on way through to goal. Could have been a red card, although there was an Everton player close by also.	Savage thinks it's a red card for Barry and he's denying a goal-scoring opportunity. Hansen also thinks Nolan is maybe going to score, and if he sends Barry off it maybe changes the game.
Mar. 3, 2014	Swansea 1 vs. Crystal palace 1	Mike Dean	Swansea's Chico is sent off for bringing down Palace striker in the box. He is deemed to be preventing a goal-scoring opportunity and is sent off. It was a foul, but I think it might have been just outside the box. Potentially a poor decision to give the penalty, but sending off was right decision.	Keown thought it was outside the box. Poor decision to give the penalty.

Table A3.3 (continued)

Date	Score	Referee	The Play	Commentators' Discussion
Mar. 8, 2014	Chelsea 4 vs. Tottenham 0	Michael Oliver	Eto'o runs onto a through ball and is brought down by Spurs goalkeeper Lloris. Play is called back for offside on Eto'o. Replay shows that Eto'o was onside and thus Lloris should have been sent off for bringing down Eto'o as the last man. Bad decision.	Fowler thinks he was onside and the keeper might have gone there.
			Eto'o is brought down in penalty area by Kaboul. He is sent off for denying a goal scoring opportunity. While the penalty is beyond doubt, I agree with the commentator that the red card seemed harsh—albeit I understand technically he was last man.	Fowler can understand why the referee has given it, and he had a good view and a good position. He didn't think he should be sent off. But it's a goal-scoring opportunity, so if it's a penalty he has to send him off. Lineker feels like he was going away from goal, so was it a goal-scoring opportunity? It looked harsh both ways to him. Murphy can see why the penalty has been given, but it's not a sending off. That kills the game. Not every time someone takes a touch in the box is it a goal-scoring chance.

Table A3.3 (continued)

Date	Score	Referee	The Play	Commentators' Discussion
Mar. 8, 2014	Crystal Palace 0 vs. Southampton 1	Howard Webb	Dejan Lovren brings down Crystal Palace striker Bolasie. Foul is given and Lovren booked. He appears to be last man (although fullback is trying to get over) and does prevent a goal-scoring opportunity, so it should have been a red.	Murphy's view when he saw it live was it's a sending-off, because he's through on goal. What you don't quite see, and we've had the luxury of seeing afterward, is the angle of the ball, etc., but instinctively he thought it was a sending off. Fowler disagrees because of the trajectory of the ball. Given the pundits are split, I've gone with my own judgment, which is that he didn't get the ball, and was last man, so it should have been a red card.
Mar. 15, 2014	Hull City 0 vs. Manchester City 2	Lee Mason	George Boyd goes down in the box under a challenge from Joe Hart. In real time it looks like it could be a penalty. Replay is quite inconclusive as it is hard to say whether Boyd left his foot in, whether there was contact or not. He appeared to wait for the contact.	Lineker thinks everyone seems to agree it was a foul. Shearer actually thinks he's down on his knees, and for him it's a dive. Given that the pundits are split, I've gone with my own judgment, which is that it was a penalty.

Table A3.3 (continued)

Date	Score	Referee	The Play	Commentators' Discussion
			Joe Hart is booked for confronting Boyd. Ithink he's lucky not to get sent off for violent conduct as he moves his head toward Boyd. In replay after the game, we also saw that Boyd spat at Hart after their heads came together. Both players should have been sent off.	Savage thinks that Hart is very fortunate here as Boyd takes a step to the side as their heads come together. Shearer thinks the FA will have a look at Boyd's spitting, but because Hart received yellow card they can't do anything about that.
Mar. 15, 2014	Aston Villa 1 vs. Chelsea 0	Chris Foy	Joe Bennett brings down Chelsea's Ramires. He appears to be last man and potentially denying a goal scoring opportunity. Mourinho and players call for red card, but only yellow is given. The replay shows there is little chance the defender can get there.	Shearer thinks the referee had a really bad day and got several big decisions wrong. He feels Bennett should have been sent off, as there is no chance the other defender can get there. Referee had poor vantage point and should have been helped by assistant.
			Willian is sent off for second yellow. Replay shows he did touch him, but was a bit harsh as Delph did appear to be already falling.	Shearer doesn't even think it's a free kick as he doesn't trip him, or pull him. He barely touches him. No way was it a yellow card.

Table A3.3 (continued)

Date	Score	Referee	The Play	Commentators' Discussion
Mar. 15, 2014	Stoke 3 vs. West Ham Utd 1	Craig Pawson	In an aerial battle with Andy Carroll, Stoke defender Muniesa appears to handle the ball amidst West Ham Utd's appeals. Replay showed he did clear it with his raised arm and it should have been a penalty to West Ham Utd.	Lineker says that they all agree it was a penalty.
Mar. 16, 2014	Manchester Utd 0 vs. Liverpool 3	Mark Clattenburg	Luis Suárez is brought down in box and appeals for a penalty. Looked like a foul in real time. Replay shows he was fouled by Fellaini. Should have been a penalty. Poor decision.	Mills thinks this was a penalty.
			Vidić brings down Sturridge in the box. Penalty given. I thought it looked like a penalty in real time, although Vidić appealed aggressively that he didn't touch Sturridge. Vidić is sent off for second yellow. Replay shows he didn't catch Sturridge. Poor decision. It shouldn't have been a penalty or a sending off.	Mills feels this was a dive. He makes it look so realistic.

Table A3.3 (continued)

Date	Score	Referee	The Play	Commentators' Discussion
			Sturridge is brought down by Carrick and appeals for penalty. Looked like a foul in real time and replay confirms it. Poor decision.	Mills feels this was a clear penalty.
Mar. 22, 2014	Chelsea 6 vs. Arsenal 0	Andre Marriner	Penalty conceded by Oxlade-Chamberlain for hand-ball. Correct decision after viewing the replay. The ball did look like it was going wide, so a yellow card would have sufficed. Moreover, as part of the same decision, the referee sends of Kieran Gibbs instead of Oxlade-Chamberlain for the hand-ball.	Shearer thinks he had a terrible day. It is a penalty, but he thinks it should only have been a yellow card because it was going wide. He feels the referee didn't see it, neither did the linesman; rather, somebody in his ear told him. Shearer believes the decision will be overturned and the right man punished.
Mar. 29, 2014	Crystal Palace 1 vs. Chelsea 0	Lee Mason	Gary Cahill brings down Crystal Palace forward in penalty box. In real time it looks like Cahill has got the ball and it is a good decision. After watching the replay though, it's clear that the Palace forward touches the ball and Cahill brings him down. Should have been a penalty.	Shearer thinks it could have been a penalty.

Table A3.3 (continued)

Date	Score	Referee	The Play	Commentators' Discussion
Apr. 5, 2014	Manchester City 4 vs. Southampton 1	Chris Foy	Samir Nasri scores for Manchester City after being set up by Silva. Southampton defenders complain about offside. Replay shows that Silva was offside and therefore the goal shouldn't stand. Bad decision by the assistant referee.	Lineker thinks it was clearly offside. Savage says it was a ridiculous decision. Only thing he can think is that linesman thought it came off a defender.
Apr. 6, 2014	West Ham Utd 1 vs. Liverpool 2	Anthony Taylor	West Ham Utd scores after Mignolet drops ball from a corner under a challenge from Carroll. Referee consults the linesman and, ultimately, the goal stands. Looked like a foul in real time and replay confirms Carroll launches into the keeper's face with his arm.	John Hartson thinks it's clearly a foul. He hits him in the face with his arm. Poor decision. Neil Lennon thinks the linesman did the right thing to flag, and the referee has made the wrong decision to overrule him.

Table A3.3 (continued)

Date	Score	Referee	The Play	Commentators' Discussion
			Penalty to Liverpool when keeper brings down Flanagan. Looked like keeper got a hand to the ball at the time, and replay confirms he did swipe the ball with his hand before contact with player. Poor decision.	Hartson thinks the keeper gets the ball first, and it's not a penalty. Lennon feels it was a penalty. He didn't think he got a strong enough touch and he brought the player down. Good decision. The view the referee has, he's got to think it's a penalty. Because the pundits are split, I've gone with my own judgment, which is that it shouldn't have been a penalty.
Apr. 13, 2014	Liverpool 3 vs. Manchester City 2	Mark Clattenburg	Sakho goes in with a stupid challenge on City's Džeko in the box. Gets nothing on the ball and I think it should have been a penalty.	Hansen thinks this was a stonewall penalty kick.
			Suárez goes down in the box under a challenge from Kompany. Looked like a penalty in real time I thought. Replay showed it was a definite penalty. Poor decision.	Hamann thinks this was a clear penalty.

Table A3.3 (continued)

Date	Score	Referee	The Play	Commentators' Discussion
			City appeals for hand-ball from Škrtel. Didn't look like much in real time. Replay shows Škrtel punched the ball. Clear penalty. Poor decision.	Hansen thinks this was a blatant hand-ball.
Apr. 19, 2014	Chelsea 1 vs. Sunderland 2	Mike Dean	Ramires elbows/backhands Larsson off the ball, seemingly right in front of the referee. It should have been a straight red card for violent conduct.	Hansen says it's certain red. It's a straight red and he can't believe the referee hasn't seen it. FA will see Ramires.
Apr. 26, 2014	Fulham 2 vs. Hull City 2	Lee Mason	Shane Long goes down in the Fulham penalty area. In real time it looks like Long has been shoved in the back as he runs through on goal. Replay shows that he was shoved—albeit he went down bit easily. It should have been a penalty and probably a red card as he was last man and stopped a goal-scoring opportunity.	Hansen thinks the defender gets on the wrong side of Shane Long and it's absolutely a penalty kick.

Table A3.3 (continued)

Date	Score	Referee	The Play	Commentators' Discussion
			Curtis Davies blatantly hand-balls in the Hull City penalty box. Strangely no Fulham players appeal. Replay confirms ball struck his high arm, and it should have been a penalty. Bad decision.	Savage thinks it was a hand-ball and it was in the box. Definite penalty.
Apr. 26, 2014	Swansea 4 vs. Aston Villa 1	Mark Clattenburg	Bony scores for Swansea amidst appeals for offside. Replay supports my initial thought that he was marginally offside. The goal shouldn't count. Poor decision.	Hansen agrees that he's marginally offside.
Apr. 26, 2014	Southampton 2 vs. Everton 0	Michael Oliver	Everton's Leon Osman goes down in the box under a challenge. It looks like a penalty, but the referee books Osman for diving. Replay shows that it was a harsh decision by the referee as there was contact, although Osman knew what he was doing.	Savage thinks it's a definite penalty. There's contact there. Ridiculous decision to book Osman for diving.
			McCarthy is pulled back on the ground by Lovren and the replay confirms that it should have been a penalty. Poor decision.	Savage thinks this is also a penalty. The player has pulled the shirt. Because the pundits are split, I've gone with my own judgment, which is that it should have been a penalty.

Table A3.3 (continued)

Date	Score	Referee	The Play	Commentators' Discussion
Apr. 26, 14	Stoke City 0 vs. Tottenham 1	Andre Marriner	Adebayor appears to elbow Stoke defender Shawcross. It's hard to tell definitively from replay, but could have been a red card for Adebayor.	Savage thinks he's very lucky to stay on the pitch. Hansen thinks it's a red card. He elbowed him. Early in the match, it might have been a very different game.
May 11, 2014	Southampton 1 vs. Manchester Utd 1	Mike Dean	Vidić appears to handle in the box. Looks like a penalty in real time. Replay confirms it's a clear hand-ball, Vidić palms it away. Should have been a penalty. Poor decision. Mike Dean can't have seen it.	Hansen thinks this is definitely a hand-ball. He's lucky to get away with it.

Appendix 4 Some Sources Relating to Sports and to Science and Technology Studies

General

The philosophical and technical analysis of this book has been updated and entirely rewritten, but while the analysis based on RINOWN is new, much of the other material is based on four published papers. We are grateful to the editors and the publishers of the following pieces to allow us to reproduce material from them. These papers are:

Collins, Harry and Evans, Robert. 2008. You cannot be serious! Public understanding of technology with special reference to "Hawk-Eye." *Public Understanding of Science* 17 (3): 283–308.

Collins, Harry, 2010. The philosophy of umpiring and introduction of decision-aid technology. *Journal of the Philosophy of Sport* 37 (2): 135–146.

Collins, Harry, and Evans, Robert. 2012. Sport-decision aids and the "CSI-effect": Why cricket uses Hawk-Eye well and tennis uses it badly. *Public Understanding of Science* 21 (8): 904–921.

Collins, Harry, and Evans, Robert. 2016. A general theory of the use of technology in sport and some consequences. In

Critical Issues in Global Sport Management, ed. Stephen Frawley and Nico Schulenkorf. Routledge.

Sports

In terms of the existing literature on sports, our focus has been very narrow, and since this is not primarily a scholarly book we will mention only a few works here in case readers want to pursue matters. This is in spite of the fact that there is a *Journal of the Philosophy of Sport* from which we have benefited. Items on sports that bear on the narrow topics discussed here and that have crossed our desk while preparing this book include the following:

Galarraga and Joyce (2012) give an account of the Galarraga incident from the perspectives of both the pitcher and the umpire.

Moskowitz and Wertheim (2011) provide a very lively and readable discussion of the way refereeing and umpiring work in US sports. Their discussion is particularly good on home advantage and why it happens.

Anderson et al. (2008) include some interesting general analyses of association football.

Helsen et al. (2006) use a psychological approach to understand referees' judgments of offside in association football.

While no arguments such as ours for the use of TV replays have been put forward before as far as we know, nor any comprehensive schema such as is found in table 7.2, we do know (as of February 2015) that certain Dutch football authorities are trying to get permission to trial TV replays and we have heard rumors that the English FA would also like to try them. Media reports on trials in the Dutch league can be found in the *Amsterdam Herald* (2014), the *National* (2015), and *World Football Insider* (2015). We hope that the analysis outlined in this book will add weight to these efforts.

Other works we found useful include the piece by Hoult (2013), who explains how Hot Spot was ready to admit its problems. We remarked in chapter 4 that track-estimator technology could do much more than estimate tracks; it could also be used to build an archive of information about how bowlers bowl, where batters strike the ball, and so forth—potentially a fascinating source of current and historical information. We remarked that this kind of thing would also be useful in tennis, and we find that it is—but that there can be a downside. Newspapers and other media outlets are ready to pay for this kind of aggregate information, and one consequence is that access to the raw data is now closely guarded. More about the use of aggregate tennis data can be found (as of February 2016), by Googling "the Australian Hawk-Eye data could serve up maps." Once more, we want to stress that we are not against the use of this kind of technology, just uses that present it as perfectly accurate.

Science Studies

What is, perhaps, the unique flavor of this book arises out of the disciplinary background of the first two-named authors. This is the field of science studies or science and technology studies. While professionals still argue violently about what the subject is and where it should be going, we take its special qualities to be the way it combines the disciplines drawn on here in a new kind of understanding of the role of science and technology in contemporary life. In this case the disciplines that had to be pulled together—and are all too rarely pulled together—are sociology, philosophy, a detailed understanding of how science works day to day, with, in this case, some ability to apply scientific thinking, and an understanding of how technology works in the real world. We think that it is this combination that defines science and technology studies.

An accessible and widely selling book by one of the authors that looks unusually closely at the practice of science is *The Golem: What You Should Know about Science* (Collins and Pinch 1993/1998). A more technical and philosophical work is Collins's (1992) *Changing Order*.

This book is mainly about sports, but it is also about more than that: it is about the public understanding of science and even about scientists' understanding of science. When we first wrote about Hawk-Eye, we discovered that even some very senior scientists were not aware of what we had pointed out. The trouble is that, in the phrase set out in *Changing Order*, in science as in love, "distance lends enchantment." That is, when scientists are not dealing with the very narrow specialism in which they are experts, they tend to fall back, like the rest of the public, into believing what they are told. Of course, they "get it" as soon as the matter is drawn to their attention—which is not necessarily true of the public—but that they don't always get it for themselves is still a surprise.

One of the things that needs constant reinforcement both in the case of professionals and the public is that models are not reality. The field of artificial intelligence has suffered for decades from this kind of elision even though it was pointed out by one of the field's heroes, Terry Winograd, as long ago as the 1980s. Winograd had built a brilliantly successful program called SHRDLU that could move things around in a world of colored blocks in response to spoken instructions. This was a breakthrough. But after Winograd had become famous, he realized that what he said he had invented was not really what he had invented at all. This was because the world in which the computer moved things was a *virtual* world—an imaginary world constructed by the computer; the computer could do marvelous things, but only in this—as Winograd called it—"microworld." It took Winograd's ability to stand back from what he had done and his readiness to criticize himself to see and explain to his artificial intelligence colleagues that a microworld is not a real world. He pointed out that success in one did not mean success in the other; things might not happen the same way in the real world because of all the extra contextual factors that are found there—and, indeed, they did not happen in the same way. That is the crucial distinction we are living through again with the track estimators. (Winograd and Flores 1987 is the source for Winograd's conversion.)

The idea that new technologies can *redefine* what counts among the public as a solution to a problem is much discussed in the sociology of technology. We draw on this when we discuss the way the bails in cricket redefine the notion of "striking the wicket" as "breaking the wicket" and when we consider how track estimators might come to redefine "in" and "out" in tennis. Key sources are:

Bijker, Hughes, and Pinch (1987): *The Social Construction of Technological Systems*. This edited collection sets out a range of different perspectives, of which the social construction of technology is one. It also contains the now classic analysis of the development of the transformation of the "penny farthing" bicycle into the modern "safety bicycle."

Bijker (1995): *Of Bicycles, Bakelites, and Bulbs*. This develops the framework set out in Bijker, Hughes and Pinch 1987 and provides a more comprehensive theory of technological innovation and adoption. The idea is that successful technologies are successful precisely because they can be integrated into social practices, and that this depends, at least in part, on how the views and interests of different social groups are mobilized to shape what counts as a "working" machine.

Notes

Introduction

1. Throughout this book we will use the term "football" to refer to the European-style game played throughout the world and sometimes known as "soccer." The American game will be referred to as "American football."

Chapter 2

1. See the "Bonus Extra" chapter on cricket near the end of the book for still more detail.

2. Some good views of cricket can be obtained (as of February 2016) on YouTube: https://www.youtube.com/watch?v=8L5PqL4ylIU; here the players in white clothes are playing the "long game," while the players in colored clothes are playing "limited-overs" cricket. (For more on this, see the "Bonus Extra" chapter on cricket.) Oddly, as far as we can see, on the European continent there is no game played that involves "the basic transaction"; there is no European continental game in which someone projects a ball with their hand and arm and someone tries to hit it with a bat. The best we can do in this book is hope that continental Europeans have seen either cricket or baseball while offering a bit of extra help here and there.

3. The laws of cricket are available at https://www.lords.org/mcc/laws-of-cricket/.

4. The rules of golf are available at http://www.randa.org/en/Rules-and -Amateur-Status.aspx; the rules of football are available at http://www .fifa.com/development/education-and-technical/referees/laws-of-the -game.html.

5. Gripper   2014,   http://www.mirror.co.uk/sport/other-sports/sochi -2014-incredible-ski-cross-796763.

Chapter 3

1. In, for example, Hawk-Eye's early analysis of a Collingwood leg before wicket decision, which, unfortunately, is no longer available on the Web.

2. Jamie Lewis put in the most time in searching the websites.

3. Rajesh   2003,   http://www.espncricinfo.com/ci/content/story/ 125938.html. S. Rajesh's quotations of the words and views of Paul Hawkins on the Cricinfo website will be referred to several times in this chapter. Rajesh works for Cricinfo and is identified on this page as its assistant editor. The relationship between Cricinfo and Hawk-Eye Innovations appears to be close. For the period between summer 2006 and summer 2007 they were both part of the same company.

4. Hawkins, quoted in Rajesh 2006, http://www.espncricinfo.com/ci/ content/story/250359.html.

5. Rajesh 2004, http://www.espncricinfo.com/ci/content/story/136222 .html. Interestingly, the sideways-on reconstruction in the case of the Collingwood lbw decision mentioned in note 1 above would suggest a frame rate of about half this.

Chapter 4

1. Gomes   2009,   http://gulfnews.com/sport/tennis/hawk-eye-leaves -nadal-and-federer-at-wits-end-1.165119.

2. Hodgkinson 2007, http://www.telegraph.co.uk/sport/tennis/wimbledon/2316731/Hawk-Eye-system-goes-under-the-microscope.html.

3. This is known as the "Peters' method" (Peters 1856). Thanks to a referee of the original paper in which these results were first presented for pointing out that a more accurate figure would be 1.253.

4. http://www.itftennis.com/media/218363/218363.pdf.

5. The full explanation was given on the Hawk-Eye website (accessed February 15, 2008; page no longer available).

6. See also Mather 2008.

Chapter 5

1. Thompson 2014, http://www.theatlantic.com/entertainment/archive/2014/09/baseball-offensive-drought-and-camera-technology/379443/.

2. For a still shot of "Howzat," see figure A.5 in the "Bonus Extra" chapter, or https://upload.wikimedia.org/wikipedia/commons/7/7c/Appeal_for_a_wicket.jpg.

3. https://en.wikipedia.org/wiki/Instant_replay_in_American_and_Canadian_football.

Chapter 6

1. This work was done by Christopher Higgins, with the method worked out between him and Harry Collins.

Conclusion

1. Hawk-Eye Innovations (n.d.), *Hawk-Eye's Accuracy & Reliability: Electronic Line Calling*, http://www.hawkeyeinnovations.co.uk/sports/tennis (accessed February 10, 2016).

Bonus Extra

1. For the historical change, see Shindler 2012, http://www
.espncricinfo.com/wisdenalmanack/content/story/573224.html. A standard scene in films featuring English rural life before World War II is the
romance between the beautiful well-born fiancée of an aristocrat who
bowls in the annual village cricket match and the rough-hewn game-
keeper or some such who smashes the ball all over the field in his untu-
tored way. In Yorkshire, it used to be said, "If you need a fast bowler,
holler down a coal mine." For a discussion of the literature of cricket,
see Tomlinson 2015, http://www.theguardian.com/books/2015/sep/02/
top-10-cricket-scenes-in-fiction.

Appendix 1

1. http://www.kitplus.com/articles/Hawk-Eye_Accuracy_and
_Believability_in_Cricket/247.html (accessed April 27, 2016).

Appendix 2

1. Collins and Pinch 2010; Postol 1991, 1992.

References

Anderson, Patrick, Peter Ayton, and Carsten Schmidt, eds. 2008. *Myths and Facts about Football: The Economics and Psychology of the World's Greatest Game*. Newcastle-upon-Tyne: Cambridge Scholars Publishing.

Bal, Baljinder, and Gaurav Dureja. 2012. Hawk Eye: A logical innovative technology use in sports for effective decision making. *Sport Science Review* 21 (1–2): 107–119.

Bijker, Wiebe E. 1995. *Of Bicycles, Bakelites, and Bulbs: Toward a Theory of Sociotechnical Change*. Cambridge, MA: MIT Press.

Bijker, Wiebe E., Thomas Parke Hughes, and Trevor J. Pinch, eds. 1987. *The Social Construction of Technological Systems New Directions in the Sociology and History of Technology*. Cambridge, MA: MIT Press.

Collins, Harry M. 1992. *Changing Order: Replication and Induction in Scientific Practice*. Chicago: University of Chicago Press.

Collins, Harry M. 2010. The philosophy of umpiring and the introduction of decision-aid technology. *Journal of the Philosophy of Sport* 37 (2): 135–146.

Collins, Harry M., and Robert Evans. 2008. You cannot be serious! Public understanding of technology with special reference to "Hawk-Eye." *Public Understanding of Science* 17 (3): 283–308.

Collins, Harry M., and Robert Evans. 2012. Sport-decision aids and the "CSI-effect": Why cricket uses Hawk-Eye well and tennis uses it badly. *Public Understanding of Science* 21 (8): 904–921.

Collins, Harry M., and Robert Evans. 2016. A general theory of the use of technology in sport and some consequences. In *Critical Issues in Global Sport Management,* ed. Stephen Frawley and Nico Schulenkorf. London: Routledge.

Collins, Harry M., and Trevor J. Pinch. 1993/1998. *The Golem: What Everyone Should Know about Science.* Cambridge: Cambridge University Press.

Collins, Harry M., and Trevor J. Pinch. 2010. *The Golem at Large: What You Should Know about Technology.* Cambridge: Cambridge University Press.

Dutch FA hopes to introduce video referees this season. 2014. *Amsterdam Herald.* http://amsterdamherald.com/index.php/rss/1231 -20140804-dutch-fa-plans-introduce-video-referees-this-season -netherlands-dutch-sport (accessed July 17, 2015).

Fischetti, Mark. 2007. In or out? *Scientific American* 297 (1): 96–97.

Football technology—Rummenigge wants goal-line tech; Dutch FA assess video refs. 2014. *World Football Insider,* May 22. http://www. worldfootballinsider.com/Story.aspx?id=36965.

Galarraga, Armando, and Jim Joyce, with Daniel Paisner. 2012. *Nobody's Perfect: Two Men, One Call, and a Game for Baseball History.* New York: Grove Press.

Gomes, Alaric. 2009. Hawk Eye leaves Federer and Nadal at wits' end. *Gulfnews.com,* May 3. http://gulfnews.com/sport/tennis/hawk-eye -leaves-nadal-and-federer-at-wits-end-1.165119.

Gower, David. 2009. Third umpire strikes back as review system improves. *Sunday Times,* Dec. 20. http://www.timesonline.co.uk/tol/ sport/columnists/david_gower/article6962837.ece.

Gripper, Ann. 2014. So-chi close!! 10 fantastic photo finishes from sporting history. *Mirror*, Feb. 20. http://www.mirror.co.uk/sport/other -sports/sochi-2014-incredible-ski-cross-796763.

Hawk-Eye Innovations. n.d. *Hawk-Eye Accuracy and Believability in Cricket.* http://www.kitplus.com/articles/Hawk-Eye_Accuracy_and _Believability_in_Cricket/247.html.

Hawk-Eye Innovations. n.d. *Hawk-Eye's Accuracy & Reliability: Electronic Line Calling.* http://www.hawkeyeinnovations.co.uk/sports/tennis.

Hawkins, Paul. 2010. Hawk-Eye accuracy and believability. (Accessed Nov. 10 2010; URL no longer available.)

Hawkins, Paul. 2015. Paul Hawkins responds to CricInfo blog: Hawk-Eye. http://www.Hawk-Eyeinnovations.co.uk/press/2015/4/21/ paul-hawkins-responds-to-cricinfo-blog-103. (Accessed July 17, 2015; URL no longer available.)

Helsen, W., B. Gilis, and M. Weston. 2006. Errors in judging "offside" in association football: Test of the optical error versus the perceptual flash-lag hypothesis. *Journal of Sports Sciences* 24 (5): 521–528.

Hodgkinson, Mark. 2007. Hawk-Eye system goes under the microscope. *Telegraph*, July 10. http://www.telegraph.co.uk/sport/ tennis/wimbledon/2316731/Hawk-Eye-system-goes-under-the -microscope.html.

Hoult, Nick. 2013. Ashes 2013: Hot Spot can miss fine edges, admits inventor ahead of third England v Australia test. http://www.telegraph .co.uk/sport/cricket/international/england/10209920/Ashes-2013-Hot -Spot-can-miss-fine-edges-admits-inventor-ahead-of-third-England-v -Australia-Test.html (accessed July 16, 2015).

ICC. 2009. *Umpire Decision Review System—a Guide.* http://www.world-a -team.com/DRS_guide.pdf.

Jonkhoff, H. C. 2003. Electronic line-calling using the TEL system. In *Tennis, Science, and Technology 2*, ed. Stuart Miller, 377–383. International Tennis Federation.

Khaled, Ali. 2015. Fifa's failure to introduce video technology is harming referee integrity, not protecting it. *National*, Mar. 1. http://www .thenational.ae/sport/football/fifas-refusal-to-introduce-video -technology-will-harm-referees-not-protect-them.

Lalor, Peter. 2010. Catching referrals "blight on cricket." *Australian*, Nov. 30. http://www.theaustralian.com.au/sport/cricket/catching-refer-rals-blight-on-cricket/story-e6frg7rx-1225962962962 (accessed July 16, 2015).

Ley, Barbara L., Natalie Jankowski, and Paul R. Brewer. 2012. Investigating CSI: Portrayals of DNA testing on a forensic crime show and their potential effects. *Public Understanding of Science* 21 (1): 51–67.

Mather, George. 2008. Perceptual uncertainty and line-call challenges in professional tennis. *Proceedings of the Royal Society B: Biological Sciences* 275 (1643): 1645–1651.

Miller, S. 2006. *Evaluation Report, Auto-Ref 5, Appendix 1—Accuracy Data*, test #37 of 120 (Aug. 2006).

Moskowitz, Tobias J., and L. Jon Wertheim. 2011. *Scorecasting: The Hidden Influences behind How Sports Are Played and Games Are Won*. New York: Three Rivers Press.

Mulkay, Michael, and Nigel Gilbert. 1984. *Opening Pandora's Box: A Sociological Analysis of Scientists' Discourse*. Cambridge: Cambridge University Press.

Owens, N. C. Harris, and C. Stennett. 2003. Hawk-Eye tennis system. In *Visual International Conference on Information Engineering 2003 (VIE 2003)*, 182–185. London: Institution of Electrical Engineers.

Peters, Christian August Friedrich. 1856. Uber Die Bestimmung Des Wahrscheimlichen Fehlers Einer Beobachtung Aus Den Abweichungen Der Beobachtungen von Ihrem Arithmetischen Mittel. *Astronomische Nachrichen* 44:30–31.

Postol, Theodore A. 1991. Lessons of the Gulf War experience with Patriot. *International Security* 16 (3): 119–171.

Postol, Theodore A. 1992. Correspondence: Patriot experience in the Gulf War. *International Security* 17 (1): 225–240.

Rajesh, S. 2003. Give Hawk-Eye a chance. *Espncricinfo.com*, Dec. 18. http://www.espncricinfo.com/ci/content/story/125938.html.

Rajesh, S. 2004. Keeping a close eye on things. *Espncricinfo.com*, Sept. 25. http://www.espncricinfo.com/ci/content/story/136222.html.

Rajesh, S. 2006. A threat or an advantage? *Espncricinfo.com*, June 13. http://www.espncricinfo.com/ci/content/story/250359.html.

Russell, J. S. 1997. The concept of a call in baseball. *Journal of the Philosophy of Sport* 24:21–37.

Shindler, Colin. 2012. The slow death of cricket's class divide. *Espncricinfo.com*, http://www.espncricinfo.com/wisdenalmanack/content/story/573224.html.

Thompson, Derek. 2014. The simple technology that accidentally ruined baseball. *Atlantic*, Sept. 4. http://www.theatlantic.com/entertainment/archive/2014/09/baseball-offensive-drought-and-camera-technology/379443/.

Tomlinson, Richard. 2015. Top 10 cricket scenes in fiction. *Guardian*, Sept. 2. http://www.theguardian.com/books/2015/sep/02/top-10-cricket-scenes-in-fiction.

Winograd, Terry, and Fernando Flores. 1987. *Understanding Computers and Cognition: A New Foundation for Design*. Reading, MA: Addison-Wesley.

Index

Inside Technology

edited by Wiebe E. Bijker, W. Bernard Carlson, and Trevor Pinch

Christopher R. Henke, *Cultivating Science, Harvesting Power: Science and Industrial Agriculture in California*

Helga Nowotny, *Insatiable Curiosity: Innovation in a Fragile Future*

Karin Bijsterveld, *Mechanical Sound: Technology, Culture, and Public Problems of Noise in the Twentieth Century*

Peter D. Norton, *Fighting Traffic: The Dawn of the Motor Age in the American City*

Joshua M. Greenberg, *From Betamax to Blockbuster: Video Stores and the Invention of Movies on Video*

Mikael Hård and Thomas J. Misa, editors, *Urban Machinery: Inside Modern European Cities*

Christine Hine, *Systematics as Cyberscience: Computers, Change, and Continuity in Science*

Wesley Shrum, Joel Genuth, and Ivan Chompalov, *Structures of Scientific Collaboration*

Shobita Parthasarathy, *Building Genetic Medicine: Breast Cancer, Technology, and the Comparative Politics of Health Care*

Kristen Haring, *Ham Radio's Technical Culture*

Atsushi Akera, *Calculating a Natural World: Scientists, Engineers, and Computers during the Rise of US Cold War Research*

Donald MacKenzie, *An Engine, Not a Camera: How Financial Models Shape Markets*

Geoffrey C. Bowker, *Memory Practices in the Sciences*

Christophe Lécuyer, *Making Silicon Valley: Innovation and the Growth of High Tech, 1930–1970*

Anique Hommels, *Unbuilding Cities: Obduracy in Urban Sociotechnical Change*

David Kaiser, editor, *Pedagogy and the Practice of Science: Historical and Contemporary Perspectives*

Charis Thompson, *Making Parents: The Ontological Choreography of Reproductive Technology*

Pablo J. Boczkowski, *Digitizing the News: Innovation in Online Newspapers*

Dominique Vinck, editor, *Everyday Engineering: An Ethnography of Design and Innovation*

Nelly Oudshoorn and Trevor Pinch, editors, *How Users Matter: The Co-Construction of Users and Technology*

Peter Keating and Alberto Cambrosio, *Biomedical Platforms: Realigning the Normal and the Pathological in Late-Twentieth-Century Medicine*

Paul Rosen, *Framing Production: Technology, Culture, and Change in the British Bicycle Industry*

Maggie Mort, *Building the Trident Network: A Study of the Enrollment of People, Knowledge, and Machines*

Donald MacKenzie, *Mechanizing Proof: Computing, Risk, and Trust*

Geoffrey C. Bowker and Susan Leigh Star, *Sorting Things Out: Classification and Its Consequences*

Charles Bazerman, *The Languages of Edison's Light*

Janet Abbate, *Inventing the Internet*

Herbert Gottweis, *Governing Molecules: The Discursive Politics of Genetic Engineering in Europe and the United States*

Kathryn Henderson, *On Line and On Paper: Visual Representation, Visual Culture, and Computer Graphics in Design Engineering*

Susanne K. Schmidt and Raymund Werle, *Coordinating Technology: Studies in the International Standardization of Telecommunications*

Marc Berg, *Rationalizing Medical Work: Decision-Support Techniques and Medical Practices*

Eda Kranakis, *Constructing a Bridge: An Exploration of Engineering Culture, Design, and Research in Nineteenth-Century France and America*

Paul N. Edwards, *The Closed World: Computers and the Politics of Discourse in Cold War America*

Donald MacKenzie, *Knowing Machines: Essays on Technical Change*

Wiebe E. Bijker, *Of Bicycles, Bakelites, and Bulbs: Toward a Theory of Sociotechnical Change*

Louis L. Bucciarelli, *Designing Engineers*

Geoffrey C. Bowker, *Science on the Run: Information Management and Industrial Geophysics at Schlumberger, 1920–1940*

Wiebe E. Bijker and John Law, editors, *Shaping Technology / Building Society: Studies in Sociotechnical Change*

Stuart Blume, *Insight and Industry: On the Dynamics of Technological Change in Medicine*

Donald MacKenzie, *Inventing Accuracy: A Historical Sociology of Nuclear Missile Guidance*

Pamela E. Mack, *Viewing the Earth: The Social Construction of the Landsat Satellite System*

H. M. Collins, *Artificial Experts: Social Knowledge and Intelligent Machines*

http://mitpress.mit.edu/books/series/inside-technology